Virology Research Progress

Advances in Viral Genomes Research

VIROLOGY RESEARCH PROGRESS

Additional books in this series can be found on Nova's website under the Series tab.

Additional e-books in this series can be found on Nova's website under the e-book tab.

GENETICS - RESEARCH AND ISSUE

Additional books in this series can be found on Nova's website under the Series tab.

Additional e-books in this series can be found on Nova's website under the e-book tab.

VIROLOGY RESEARCH PROGRESS

ADVANCES IN VIRAL GENOMES RESEARCH

JOHN A. BORRELLI
AND
YLENIA D. GIANNINI
EDITORS

New York

For permission to use material from this book please contact us:
Telephone 631-231-7269; Fax 631-231-8175
Web Site: http://www.novapublishers.com

NOTICE TO THE READER

Additional color graphics may be available in the e-book version of this book.

Library of Congress Cataloging-in-Publication Data

ISBN: 978-1-62808-723-9

LCCN: 2013945781

Published by Nova Science Publishers, Inc. † New York

Contents

Preface

Viral diseases have an important impact on public health worldwide. New genomic technologies are providing infectious disease scientists with a unique ability to study at the genetic level those viruses that cause disease and the interactions they have with infected hosts. In this book, the authors present new research in viral genomes. Topics include improvements in HSV-1 derived amplicon vectors for gene transfer; viral genome research in papillomavirus; the synthetic synthesis of viral genomes; and a novel bioinformatic method to analyze more than 10,000 influenza virus strains simultaneously.

Chapter 1 – Viral vectors engineered to carry transgenic sequences can be delivered into discrete tissues or anatomical structures to express specific transgenes into the transduced cells. Therefore, they are useful tools to produce specific, transient and localized knockout, knockdown, ectopic expression or overexpression of a gene, leading to the possibility of analyzing both *in vitro* and *in vivo* molecular basis of relevant functions. Replication-incompetent helper-dependent amplicon vectors, derived from herpes simplex virus type-1 (HSV-1) are devoid of viral genes. Thus, these vectors have great advantages as tools for *in vitro* and *in vivo* gene transfer and, in particular (i) minimal toxicity or induction of adaptive immune responses, and (ii) large transgene capacity, being able to carry up to 150 kbp of foreign DNA. In addition, these vectors have (iii) widespread cellular tropism: amplicons can experimentally infect several cell types, either quiescent or not, though naturally HSV-1 infects mainly neurons and epithelial cells, and (iv) absence of insertional mutagenesis, since the viral genome does not integrate into the host cell genome. These vectors have been used both on basic and applied research, and they have revealed as most suitable tools to study complex functions involving the nervous system, such as anxiety, sexual behavior, learning and memory. In addition, amplicon vectors are being used for the development of new experimental gene therapy approaches, both for inherited and acquired diseases affecting the nervous system, including neurodegenerative diseases. Although several technological improvements have been achieved in the last decade, some difficulties regarding these appealing vectors remain still unresolved, such as the inability to generate large amounts of high-titer fully helper-free vectors and the fact that expression from the transgenic sequence delivered by the vectors is generally unstable, often leading to a complete silencing of expression after a few weeks. To overcome these obstacles and to improve these vectors, the authors have recently modified (a) the amplicon genome, in order to fully delete bacterial sequences and (b) developed novel complementing cell lines, in order to improve helper-free vector production and to render amplicon stocks compatible with clinical trials. In this review article they briefly review data supporting the potential of HSV-1-based amplicon vector model for gene delivery in primary cultures of neural cells and into the brain of living animals.

Chapter 2 – Papillomaviruses (PVs) infect the epithelium of amniotes, where they can cause tumours or persist asymptomatically. PVs are classified in the *Papillomaviridae* family, that contains 29 genera of PVsisolated from humans (120 types), non-human mammals, birds and reptiles (69 types). PVs have circular double-stranded DNA genomes approximately 8 kb in size and typically contain eight genes. Studies aiming the identification of PVs genomes use techniques such as PCR with consensus primers, rolling circle amplification and metagenomic methods. Advances in papillomaviral genome research have allowed the knowledge of PVs diversity and evolution of the poorly known PVs genera types, revealing that there is still a limited understanding of PVs diversity. Particularly, recent studies in Bovine Papillomavirus (BPV) have shown the identification of novel BPV types and several putative new virus types in cattle. This chapter will show new contributions in PVs genome studies.

Chapter 3 – Viral particles are important tools in Molecular Biology, acting as carriers of genetic material, immunogenic antigens, adjuvants, or even directly combating antibiotic-resistant microorganisms and their biofilms in hospital and industrial environments. However, the efficient use of these particles requires extensive knowledge about their characteristics and components, including those involved in their regulatory mechanisms of genome transcription and protein synthesis. The exploration of this knowledge becomes a challenge especially for scientists analyzing the virions in their natural environment, due to their interactions with the complex and diverse types of biological systems, which directly influence the regulation of the infective cycles. Thus, knowledge about viral genes, their function, organization, and modulation, beyond the comprehension of the viral components as parts of complex systems, consist of the main hurdles for the controlled and predictable handling and use of these particles. For this, the technology of Synthetic Synthesis of viral genomes is distinguished from traditional genetic engineering through the use of modularity and standardization to construct proof-of-principle systems and allow generalized circuits designs to be applied to different scenarios. This new technology is made possible thanks to advances in many areas of science, from the use of restriction enzymes until the development of techniques of genomic synthesis and sequencing, like the 454 Roche, Illumina, and SOLiD systems. This technology is becoming increasingly a multidisciplinary tool used in the investigations about these complex systems, as well in the engineering of new particles for the optimization of diverse viral functions and alterations in their infectivity and affinity, or even in the development of completely new organisms and features without the need for a template. Nevertheless, advances in this technology are still limited by the lack of dynamic techniques for monitoring biological systems and efficient and standardized circuits. Here, the authors summarize the major characteristics of viral genomes, their organization and gene modulation, and highlight the main aspects of Synthetic Biology applied to viral genomes, as its main techniques and applications.

Chapter 4 – With increase of microbial and viral genome sequence data obtained from high-throughput DNA sequencers, novel tools are needed for comprehensive analyses of the big sequences data. An unsupervised neural network algorithm, Self-Organizing Map (SOM), is an effective tool for clustering and visualizing high-dimensional complex data on a single map. The authors previously modified the conventional SOM for genome informatics on the basis of batch-learning SOM (BLSOM), by making the learning process and resulting map independent of the order of data input. Influenza virus is one of zoonotic viruses and shows clear host tropism. Important issues for bioinformatics studies of influenza viruses are prediction of genomic sequence changes in the near future and surveillance of potentially hazardous strains.

To characterize sequence changes of influenza virus genomes after invasion into humans from other animal hosts and to study molecular evolutionary processes of their host adaptation, the authors have constructed BLSOMs for oligonucleotide, codon, amino-acid, and peptide compositions in all genome sequences of influenza A and B viruses and found clear host-dependent clustering (self-organization) of the sequences.

Viruses isolated from humans and birds differ in mononucleotide composition from each other. In addition, host-dependent oligonucleotide and peptide compositions that cannot be explained with the host-dependent mononucleotide composition are revealed by these BLSOMs. Retrospective time-dependent directional changes of oligonucleotide compositions, which are visualized for human strains on BLSOMs, can provide predictive information about sequence changes of the newly invaded viruses from other animal sources.

Basing on this host-dependent oligonucleotide composition, the authors have proposed a strategy for prediction of directional changes of virus sequences and for surveillance of potentially hazardous strains when introduced into human populations from nonhuman sources. Millions of genomic sequences from infectious microbes and viruses will become available in the near future because of their medical importance, and BLSOM can characterize such big data easily and support efficient knowledge discovery.

In: Advances in Viral Genomes Research ISBN: 978-1-62808-723-9
Editors: J. Borrelli and Y. Giannini

Chapter 1

Improvements in HSV-1-Derived Amplicon Vectors for Gene Transfer

Matias E. Melendez[1,2*], Alejandra I. Aguirre[3,*], Maria V. Baez[3], Carlos A. Bueno[1,4], Anna Salvetti[5], Diana A. Jerusalinsky[3,†] and Alberto L. Epstein[1,5,†]

[1]Centre de Génétique et de Physiologie Moléculaire et Cellulaire, CNRS UMR5534, Villeurbanne, France

[2]Molecular Oncology Research Center, Barretos Cancer Hospital, SP, Brazil

[3]Instituto de Biología Celular y Neurociencia "Prof. Eduardo De Robertis", CONICET-UBA, Facultad de Medicina, Universidad de Buenos Aires. (1121) Ciudad de Buenos Aires, Argentina

[4]Laboratorio de Virología, Departamento de Química Biológica. Facultad de Ciencias Exactas y Naturales, Universidad de Buenos Aires, Argentina

[5]Centre International de Recherche en Infectiologie (CIRI) INSERM U1111 – CNRS UMR5308 Université Lyon 1, Lyon, France

---

[*] Equivalent contribution to this work.

[†] Equal Contribution to This Work and Corresponding authors.

Alberto L. Epstein alberto.epstein@ens-lyon.fr; CIRI INSERM U1111 – CNRS UMR5308. Université Lyon 1, ENS de Lyon. 46 Allée d'Italie. 69364 LYON cedex 07, France.

Diana A. Jerusalinsky djerusal@Gmail.com; IBCN - School Medicine. University of Buenos Aires. 2155 Paraguay street, 3rd floor. (1121) City of Bs. As., Argentina.

Abstract

Viral vectors engineered to carry transgenic sequences can be delivered into discrete tissues or anatomical structures to express specific transgenes into the transduced cells. Therefore, they are useful tools to produce specific, transient and localized knockout, knockdown, ectopic expression or overexpression of a gene, leading to the possibility of analyzing both *in vitro* and *in vivo* molecular basis of relevant functions. Replication-incompetent helper-dependent amplicon vectors, derived from herpes simplex virus type-1 (HSV-1) are devoid of viral genes. Thus, these vectors have great advantages as tools for *in vitro* and *in vivo* gene transfer and, in particular (i) minimal toxicity or induction of adaptive immune responses, and (ii) large transgene capacity, being able to carry up to 150 kbp of foreign DNA. In addition, these vectors have (iii) widespread cellular tropism: amplicons can experimentally infect several cell types, either quiescent or not, though naturally HSV-1 infects mainly neurons and epithelial cells, and (iv) absence of insertional mutagenesis, since the viral genome does not integrate into the host cell genome. These vectors have been used both on basic and applied research, and they have revealed as most suitable tools to study complex functions involving the nervous system, such as anxiety, sexual behavior, learning and memory. In addition, amplicon vectors are being used for the development of new experimental gene therapy approaches, both for inherited and acquired diseases affecting the nervous system, including neurodegenerative diseases. Although several technological improvements have been achieved in the last decade, some difficulties regarding these appealing vectors remain still unresolved, such as the inability to generate large amounts of high-titer fully helper-free vectors and the fact that expression from the transgenic sequence delivered by the vectors is generally unstable, often leading to a complete silencing of expression after a few weeks. To overcome these obstacles and to improve these vectors, we have recently modified (a) the amplicon genome, in order to fully delete bacterial sequences and (b) developed novel complementing cell lines, in order to improve helper-free vector production and to render amplicon stocks compatible with clinical trials. In this review article we briefly review data supporting the potential of HSV-1-based amplicon vector model for gene delivery in primary cultures of neural cells and into the brain of living animals.

Keywords: HSV-1-derived vectors, *Amplicons, *and nervous system, *and neurodegenerative disorders, *and cancer, *and vaccines, Experimental gene therapy

HSV-1 Background

HSV-1 Epidemiology and Clinical Features

HSV-1 is one of the most common human pathogens, infecting 40–80% of people world-wide. Primary, productive infections of HSV-1 typically occur in regions of the orofacial mucosa, causing a mild gingivo-stomatitis. The virus particles produced during this primary infection enter the peripheral sensory neurons, where the virus genome often remains into a latent for the lifetime of the host. Reactivation, occurring generally following stress, usually leads to recurrent lesions in the vicinity of the primary infected area and are typically limited to cold sores of the mouth. Although recrudescent herpes may affect any site along the involved sensory division of the fifth cranial nerve, the most frequent location is the muco-cutaneous junction of the lips [1]. Classic lesions are characterized by virally induced epithelial damage and go through different clinical stages. Initially, a transient cluster of micro-vesicles appears at the site of recrudescence and breaks open to form irregular, superficial erosions that crust over and heal without scarring over a period of 1–2 weeks. Shedding of the virus is often present for several days after resolution of clinical signs and symptoms [1]. HSV-1 infection can also cause ocular herpes and, much more infrequently, it can cause encephalitis. In neonates and in immune-depressed individuals HSV-1 can cause severe disseminated infection with neurological impairment and high mortality [2].

HSV-1 Virion Structure

The architecture of this enveloped double-stranded DNA virus, of 220 nm in diameter, is highly complex [3]. The mature HSV-1 virion (Fig. 1) consists of the following four components:

- A core of double-stranded (ds) DNA
- An icosadeltahedral capsid
- The tegument, which is a layer of proteins located between the envelope and the underlying capsid; these proteins play a critical role in virion morphogenesis, and they also help the virus to take the control of the expression machinery of the cell very early following infection

- A lipid envelope from cellular origin, where viral glycoproteins and other membrane proteins are embedded, several of which are involved in receptor-mediated cellular entry.

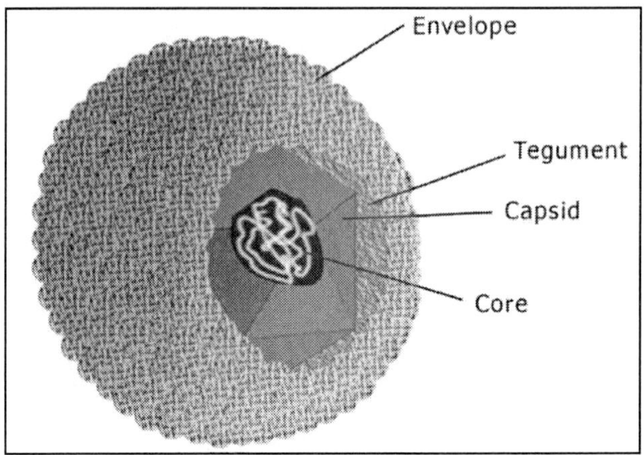

Figure 1. HSV-1 Virion.

HSV-1 Genome

The linear 152-kbp dsDNA virus genome (Fig. 2) is densely packaged within the capsid cavity [4] and is devoid of histone proteins at this stage [5]. This genome is arranged as long (UL) and short (US) unique segments, of 126 and 26 kbp, respectively. The UL and US segments are flanked by repeated sequences, designated ab, b'a' a'c' and ca (or TRL, IRL, IRS and TRS, for Terminal Repeat L, Internal Repeat L, Internal Repeat S and Terminal repeat S) [6].

The HSV-1 genome, which is 68% G+C rich, has been fully sequenced [7-9]. At least eighty-four viral genes are encoded, and these may be divided into essential and non-essential genes, according to whether their expression is necessary or not for viral growth in permissive cultured cells. HSV-1 genes do not contain intron sequences, with the exception of those encoding ICP0, ICP22, ICP47, UL15 and LAT. Nonessential genes often encode functions that are important for specific virus-host interaction *in vivo*, such as immune evasion, replication in non-dividing cells or shutdown of host protein synthesis. Approximately half of the viral genes have been shown to be dispensable for replication of the virus in cultured cells, and thus, these genes could be replaced by exogenous genetic material, which has been the premise for the development of HSV-1-based vectors for gene therapy [10]. In contrast, most genes involved in virus entry, DNA replication, capsid assembly and DNA packaging are essential.

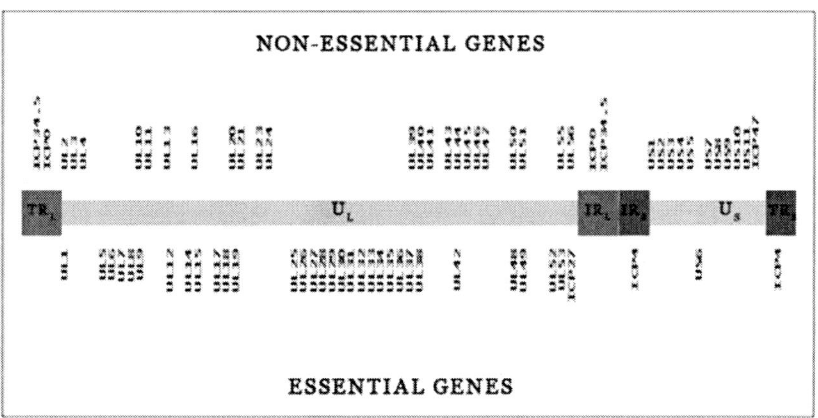

Figure 2. HSV-1 genome.

The HSV-1 genome contains two *cis*-acting sequences that are essential for virus multiplication: the viral DNA replication origins and the cleavage/packaging signals. HSV-1 harbors three lytic origins of replication, with two located within the repeated segments surrounding US (oriS) and one in the UL segment (oriL) [11]. Mutant viruses lacking either oriL or both copies of oriS are replication competent, suggesting that all origins are functionally competent [6]. The oriS is located within the shared promoter regulatory region between the divergently transcribed immediate-early genes ICP4 and ICP22/ICP47, whereas oriL lies in the regulatory region between two divergently transcribed early genes that are essential for viral DNA replication: the major DNA-binding protein and the HSV-1 DNA polymerase [12].

The cleavage/packaging signals of the HSV-1 genome are located within the *a* sequences, generally found in tandem repeats at both ends of the genome, as well as at the L/S junction [13, 14]. In different HSV-1 strains, the *a* sequence ranges from 250 to 500 bp. An approximately 200-bp fragment (Uc-DR1-Ub) spanning the junction between tandem *a* sequences has been shown to contain all the essential *cis*-acting sequences necessary for DNA cleavage and packaging [15]. The specific signals for DNA cleavage and packaging, termed pac1 and pac2, are located within the Ub and Uc regions, respectively [14-16].

HSV-1-Derived Vectors

The improvement of methods for efficient delivery of genetic material in mammalian cells has been a major objective of molecular and cellular biology, gene therapy and vaccine development during the last 30 years and is still a subject of increasing interest. Viral-derived vectors are the most promising gene transfer tools as viruses have naturally evolved as molecular carriers to specific tissues. As already quoted, HSV-1 is a naturally epithelial-neurotropic virus, highly adapted to the nervous system environment of the host organism, and neurons and glia are thus the most common target cells for HSV-1-derived vectors. Coding more than 80 genes, HSV-1 is a complex virus, which can be engineered to exquisitely design viral vectors for fundamental research and gene therapy of neurological disorders. In general terms, HSV-1 possesses several features that make it especially interesting as a vector:

- Although essentially neurotropic, under experimental conditions HSV-1 has a broad host cell range, due to the fact that its cellular receptors are widely distributed. In addition, HSV-1 infects and replicates in cells from different mammals, easily allowing preclinical evaluation.
- It can infect dividing and non-dividing cells.
- It generally causes nonthreatening diseases.
- It has a very large transgene capacity, allowing to deliver multiple or large transgenes (approximately half of the genome is nonessential in cultured cells and can be deleted without compromising viral replication, allowing to lodge considerable amounts of transgenic DNA). Moreover, HSV-1-derived amplicon vectors are devoid of almost the totality of their genome (i.e., 152-kbp DNA), which can therefore be replaced by exogenous DNA.
- Safe replication-defective HSV-1 vectors may be prepared to high titers.
- The ability to establish a latent state in neurons may be very useful for stable long-term expression of therapeutic transgenes.
- HSV-1 genome does not integrate into host chromosomes, remaining episomal and reducing the risk of insertional mutagenesis.

To exploit the versatility of HSV-1 for gene transfer, three types of HSV-1-based vectors have been developed: (*i*) attenuated replication-competent vectors, (*ii*) replication-defective recombinant vectors, and (*iii*) amplicon vectors. Attenuated replication-competent HSV-1 vectors can replicate only in certain cell types and tissues *in vivo* due to deletion of non-essential viral genes. Such vectors have been typically used for the development of tumor therapies, and are usually referred to as oncolytic HSV-1 vectors. However, other uses of attenuated vectors are possible, for example to deliver genes to the peripheral sensory ganglia following infection at peripheral tissues. Replication-defective recombinant HSV-1 vectors are devoid of one or more viral genes that are essential for lytic replication (such as the immediate early genes ICP4 and ICP27), but they can retain their ability to establish latency [17, 18]. The production of these vectors requires complementing cell lines expressing the deleted viral genes. This type of vectors can be used as genetic vaccines or for gene therapy of neurological disorders.

Lastly, Spaete and Frenkel [19] reported the natural generation of fully defective HSV-1 genomes, in which almost all the transacting virus genes were absent. These defective genomes were formed mainly by repeats of the two non-coding HSV-1 sequences, i.e., the origins of DNA replication and the cleavage/packaging signals. Moreover, they have showed that these defective genomes could be replicated and packaged into particles in the presence of a helper virus that supplied all the virus functions *in trans*, thus demonstrating that the origins of virus replication and the cleavage/packaging signals were the only two *cis*-acting sequences required for replication and packaging of a defective virus genome. These *cis*-acting sequences from the HSV-1 genome were then incorporated into a standard bacterial plasmid, which they have termed the amplicon plasmid [19], and they demonstrated that the amplicon plasmid could also be replicated and packaged into viral particles in the presence of a helper system, thus generating what they have termed amplicon vectors. This chapter will be devoted to the biology and applications of this very appealing type of vectors.

Amplicon Vectors

A conventional HSV-1 amplicon plasmid consists of (*i*) a plasmid backbone harboring a bacterial origin of DNA replication and an antibiotic resistance gene for propagation in bacteria, (*ii*) an HSV-1 origin of DNA replication (generally oriS), (*iii*) an HSV-1 cleavage/packaging sequence (pac or *a*), ***(iv)*** a transcription unit expressing a reporter gene, generally encoding the green fluorescence protein (GFP), which is useful for identifying the amplicon-infected cells and to facilitate the titration of amplicon stocks, and (***v***), a multiple cloning site (MCS) for introducing the transgene(s) of interest (Fig. 3). This basic structure of amplicon plasmids has remained almost unchanged from the beginning [19].

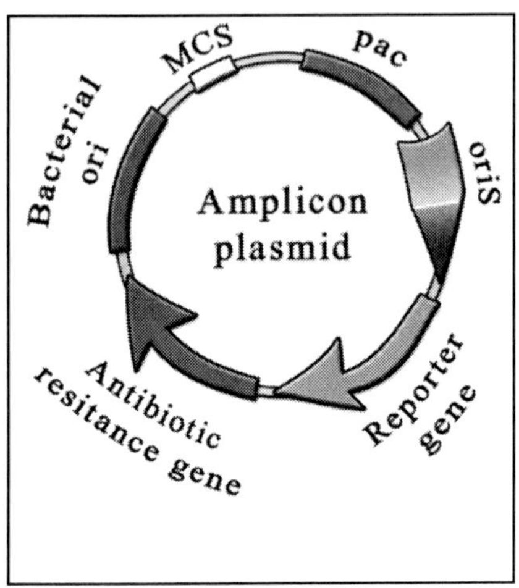

Figure 3. Structure of a typical amplicon plasmid.

Heterologous transcription units, either alone or in combination, can be cloned into amplicon plasmids using conventional molecular cloning techniques; the resulting construct is then packaged into viral particles to be then used for transduction of cells or tissues. From the structural point of view, amplicon vectors are virus particles structurally and immunological identical to wild-type HSV-1 particles, but bearing, instead of the virus genome, a concatemeric form of DNA amplified in a head-to-tail arrangement, which derives from the amplicon plasmid (Fig. 4).

Amplicon vectors have been used both on basic and applied research, and they have revealed as most suitable tools to study complex functions involving the nervous system, such as anxiety, sexual behaviour, learning and memory. In addition, amplicon vectors are being used for the development of new experimental gene therapy approaches (both for inherited and acquired diseases affecting the nervous system, including neurodegenerative diseases) [20-23], vaccine development [24, 25], hybrid viral vectors (such as adeno-associated and lentivirus viruses) [26, 27], human artificial chromosome technologies [28, 29] and basic research. It is not possible to fully describe these applications in the short extent of this manuscript. For a more detailed description of amplicon vector applications, please refer to Cuchet *et al.* (2007) [30], Epstein (2009) [31],Fraefel (2007) [32], Manservigi *et al.* (2011) [33] and de Silva and Bowers (2009) [10].

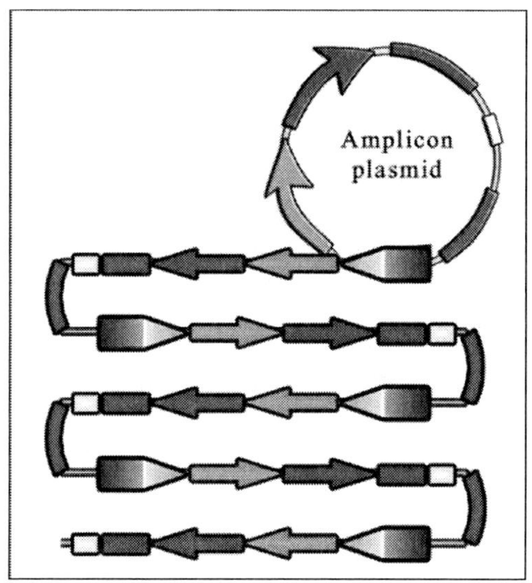

Figure 4. Concatemeric amplicon DNA replication.

Amplicon Vector Properties

The amplicon system has several properties that make it a reliable and efficient method for gene transfer.

1. The two HSV-1 elements required to support replication and packaging into virions, i.e., the oriS and *a* sequences, are smaller than 1-kbp. Since the HSV-1 particle can package up to 153-kbp, this means that the amplicon particle has the potential to accommodate very large fragments of foreign DNA, including multiple and/or large transgenes or large cell type-specific regulatory sequences. This is the most remarkable feature of amplicon vectors, as there is no other available viral vector system displaying the capacity to deliver such a

large amount of foreign DNA (~150-kbp) to the nuclear environment of mammalian cells.

2. Depending on the size of the amplicon plasmid, several copies of the transgene can be packaged into a single vector particle [34], thereby increasing the transgene dose per infected cell. This happens because the amplicon genome is replicated via a mono-directional rolling circle-like mechanism, therefore generating long concatemers composed of tandem repeats of the amplicon plasmid, and because each HSV-1 amplicon particle always cleaves and packages a ~150-kbp DNA molecule, which represents the HSV-1 particle packaging capacity [35].
3. The HSV amplicon is not an 'alive' virus but an inert particle, bearing an inherently safe *in vivo* profile. Although amplicon genomes carry no virus genes and consequently do not induce synthesis of viral proteins, the approximately 40 different HSV-1 structural proteins that shape the HSV-1 virion (which are the same in the amplicon virion) are delivered into the cell during infection and can trigger cell signaling and cellular responses, probably having a transient impact on cell homeostasis and cellular gene expression. Nevertheless, these proteins soon disappear and the cells can resume their normal functions, including the ability to divide and to respond to physiological stimuli. Furthermore, the absence of viral genes in the amplicon genome strongly reduces the risk of reactivation, complementation or recombination with latent or resident HSV-1 genomes.
4. As already quoted, after primary HSV-1 infection of epithelial cells, the virus particles can enter sensory nerve endings and the capsid is moved via retrograde axonal transport to the neuronal cell bodies of the sensory ganglia, where the virus genome may establish a latent infection. Most of these mechanisms are certainly reproduced by amplicon vectors, since the mature amplicon particle is identical to that of the helper HSV-1 and they will thus behave as normal virion particles. However, it is not yet clear if the amplicon genome can establish a latency-like long-term expression in sensory neurons.
5. Once packaged into viral particles, the amplicon vector retains the ability to infect numerous cell types, and its genome is maintained in an episomal state within the nucleus of the infected cell. Due to the absence of viral gene expression, the amplicon is replication-defective, and its episomal existence results in stable maintenance in

post-mitotic cells, but leads to unequal segregation in mitotically active cells. Since it does not integrate into the host cell genome, the conventional amplicon does not lead to insertional mutagenesis, thus increasing its safety profile as a gene therapy vector.

Amplicon Production Systems

Production of amplicon vectors involves the coordinated action of more than 50 viral proteins, required to allow replication and packaging of the amplicon plasmid into fully infectious virus particles. Amplicon plasmids are therefore dependent on helper HSV-1 virus proteins, and the genes encoding these proteins could in principle be supplied by a helper HSV-1, or by HSV-1 viral DNA.

The most important limitation for the use of amplicons in gene therapy trials is the difficulty to produce large, high-titer stocks of vector particles free of helper virus contamination. At present, there are two major methods used for producing amplicon vectors with low or even without HSV-1-helper-contamination (Fig. 5): one is based on transfection of the amplicon plasmid with HSV-1 helper genome (either as bacterial artificial chromosome (BAC) or as cosmid-based packaging systems), and the other, is based on transfection of the amplicon plasmid in cells expressing Cre recombinase, followed by infection using a defective helper HSV-1 carrying loxP sequences surrounding the packaging signals.

Helper DNA-Based Packaging Methods

DNA-based packaging methods for amplicon production are carried out by co-transfecting amplicon plasmids with HSV-1 genomes lacking their packaging signals. This system, initially derived from a set of five overlapping cosmids, each one carrying a large fragment of the virus genome, which allows the reconstitution of a full virus genome by homologous recombination, in cells co-transfected with the full cosmid set [36]. By deleting the *a* sequences from the two cosmids carrying them, this system generates a virus genome that is able to express all the required transacting functions but which cannot itself be packaged into HSV-1 particles. The co-transfection of amplicon plasmids with the modified set of cosmids allows therefore the production amplicon vectors. Using this system, however, amplicons are produced in relative low amounts (10^5-10^6 TU/mL) [37], and the vector stocks can be barely contaminated with helper virus particles.

This approach was later simplified and significantly improved by cloning the entire HSV-1 genome, only devoid of *a* signals, into a bacterial artificial chromosome (BAC), thus generating a BAC-HSV-1) [38, 39]. In the last version of this system, the gene encoding the essential ICP27 protein was further deleted from the BAC-HSV-1 genome, which was then increased in size by adding non-coding DNA to further reduce the probability of being packed into newly assembled HSV-1 particles [39, 40]. The ICP27 protein is supplied *in trans,* both by a plasmid and by complementing cell lines expressing this protein (Fig. 5A). This system gave rise to the production of entirely helper-free amplicon stocks for the first time. Vector titers obtained from helper virus-free amplicon packaging can range from 10^7 to 10^8 TU/ml, but the total amount of particles is somewhat limited by the fact that vector stocks produced in this way cannot be serially passaged.

Helper Virus-Based Packaging Methods

Historically, the first method to produce amplicon stocks was based in the transfection of cells with the amplicon plasmid, followed by super-infection with a helper HSV-1 virus, to supply *in trans* the necessary viral functions. One advantage of this system was that the vector/helper stock thus produced could be serially passaged, to produce as many amplicon particles as required. Although this method allows to produce large amounts of amplicon vectors, its major limitation is that it leads to a mixed population of particles, highly contaminated with HSV-1 helper particles, which induces strong cytotoxicity and immune responses upon infection of target cells or organisms, thus impeding their use in gene therapy and even in many fundamental studies. Due to the identical physical properties of both amplicon and HSV-1 helper virus particles, a selective purification of amplicon particles by physical treatment is not feasible.

Therefore, different strategies have been successively developed in order to reduce the toxicity of the helper-contaminated amplicon stocks, firstly by modifying the helper virus genome in order to limit its toxicity, and secondly, by limiting the production of contaminant particles. The first approach to improve this virus-based method was the use of a thermosensitive HSV-1 as helper [41]. Subsequently, defective helper HSV-1 viruses, carrying deletions in immediate early genes encoding one essential protein [42-45] were developed. However, and despite obtaining relatively high titers, these amplicon stocks were still contaminated with large amounts of defective HSV-1 helper particles, resulting in significant cytotoxicity and inflammation.

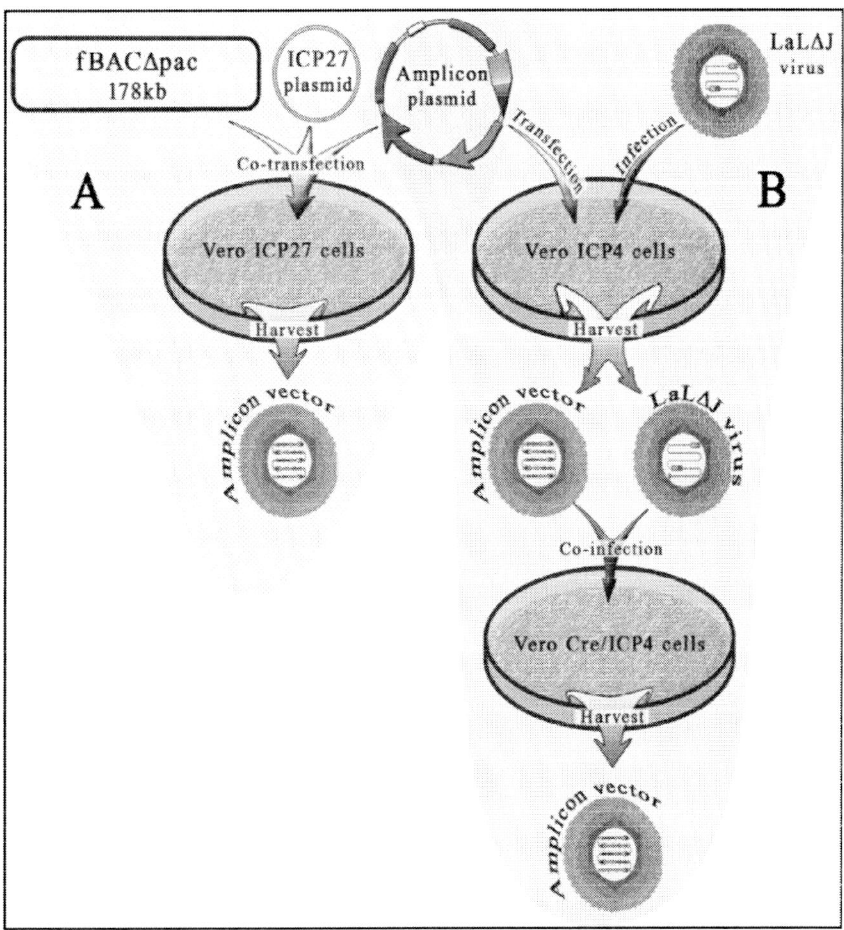

Figure 5. Amplicon production systems. (*A*) DNA-based packaging system: Vero ICP27-expressing cells are co-transfected with the amplicon plasmid, the fBACΔpac BAC (which carries a non-packageable HSV-1 genome) and an ICP27-expressing plasmid, and amplicon vectors are harvested from cells 2 or 3 days post-transfection. (*B*) Helper virus-based packaging system: Vero ICP4-expressing cells are transfected with an amplicon plasmid and super-infected with the HSV-1 *LaLΔJ* helper virus. At two days post-infection, the mixed population of viral particles (amplicons and helper particles) is used to infect ICP4/Cre recombinase-expressing cells. After 2 days, amplicon vectors, only barely contaminated with defective helper particles, are harvested from infected cells.

LaL and LaLΔJ HSV-1 Helper System

More recently, an attractive approach to produce amplicon vectors using the helper virus-based system, was developed to avoid or to limit helper genome packaging while producing high amounts of amplicon stocks. This system is based on the deletion of *a* sequences from the HSV-1 helper genomes by a Cre/loxP-based site-specific recombination, in order to abolish their packaging in the cells that are producing the amplicon vectors. The first of these helper systems, named HSV-1 *LaL* (for lox-*a*-lox) helper, carried a unique and ectopic *a* sequence flanked by two parallel loxP sites, located in the gC locus [46, 47]. This virus is therefore Cre-sensitive and cannot, in principle, be packaged in Cre-expressing cells due to deletion of the floxed *a* sequence. Nevertheless, some helper genomes could escape the action of the Cre recombinase, allowing the production of some helper particles, which are replication-competent and able to spread.

To further improve this helper system, the two genes surrounding the *a* sequence, encoding the virulence factor ICP34.5 and the essential protein ICP4, were deleted from the *LaL* genome, generating the *LaLΔJ* helper virus [48]. Although the amplicon stocks prepared with this helper, in a cell line trans-complementing both Cre and ICP4 proteins (Fig. 5B) still contain a very small amount of contaminating helper particles, the replication-incompetent *LaLΔJ* helper cannot spread upon infection of target cells or tissues. Use of the HSV-1 *LaLΔJ* helper virus generally results in the production of large stocks of amplicon vectors with titers often exceeding $2x10^8$ TU/mL, and only barely contaminated (about 1.0%) with defective, non-pathogenic helper particles. Nevertheless, the presence of contaminating helper virus, even at very low levels, could still be a limitation for the use of amplicon stocks in some gene therapy applications. Therefore, the approaches to produce amplicon vectors should still be improved in order to produce high amounts of high-titer of entirely non-toxic amplicon stocks.

Further Improvements of Amplicon Vectors Technology

Although several technological improvements for amplicon vectors production have been thus developed in the last decade, several difficulties remain still unresolved. These include: (***i***) cytopathic effects and immune responses induced by gene expression from potential helper virus contaminants; (***ii***) potential reversion of helper virus contaminants to the wild type HSV phenotype; (*iii*) potential interactions with residing HSV-1, eventually leading to complementation, reactivation and recombination; (*iv*) the inability of the vector genome to be maintained in proliferating cells; (*v*) the presence of bacterial DNA sequences, which can lead to inflammatory responses and to the silencing of transgene expression, and (*vi*) the already quoted inability to generate large amounts of high-titer vectors fully free of contaminant helper particles. To overcome some of these obstacles and to improve these vectors, we have recently introduced two modifications to the amplicon vector production system, corresponding to: (*i*) the amplicon genome, in order to fully delete bacterial sequences and (*ii*) the nature of the complementing cell lines where the amplicons are produced, in order to improve helper-free vector production. We briefly describe in the following paragraphs our ongoing program to improve the amplicon methodology.

Elimination of Bacterial DNA from the Amplicon Genome

It is known that the presence of bacterial sequences in transduced plasmids can induce silencing of the transgene cassette, as well as innate and inflammatory reactions. In addition, bacterial sequences are outlaw in gene therapy protocols. In the case of amplicons, it was recently demonstrated that the presence of bacterial sequences in the amplicon plasmid (and thus, also present in the amplicon vector genome) cause inflammatory responses and rapid transgene silencing, by forming inactive chromatin [49]. In this regard, it has been shown that infection with amplicon vectors lacking bacterial sequences (minicircle amplicon vectors) induced approximately 20-fold higher transgene expression due to transcriptional enhancement [49]. In addition, nude mice injected with minicircle amplicon vectors exhibited 10-fold higher luciferase expression than mice injected with conventional amplicons, detectable up to at least 28 days post-infection [49]. Elimination of these bacterial DNA sequences from the amplicon genome is therefore critical if the vectors will be used in gene therapy.

Our efforts to eliminate bacterial DNA from the amplicon genome are based on the modification of the amplicon plasmid. We have recently generated a new amplicon plasmid, named pOPNE (Fig. 6), which possesses a typical plasmidic backbone with a bacterial origin of DNA replication (*E. coli* ori) and an antibiotic resistance gene (AmpR). This plasmid also contains an amplicon cassette surrounded by two loxP (L) sites in parallel orientation. As usual, the amplicon cassette carries one HSV-1 oriS and one pac (or *a*) signal.

Note that the oriS sequence is linked to the HSV-1 IE4/5 promoter (IE4/5), very close to one of the loxP sites. pOPNE also contains a Neomycin /Kanamycin resistance gene (Neo) bordered by two FRT sites (F). This FRT-bordered cassette will serve as an acceptor locus for the introduction of transgenes through FRT-flipase site-specific recombination. Finally, the amplicon cassette also possesses a promoterless EGFP (enhanced-GFP) reporter gene, juxtaposed to the second loxP site.

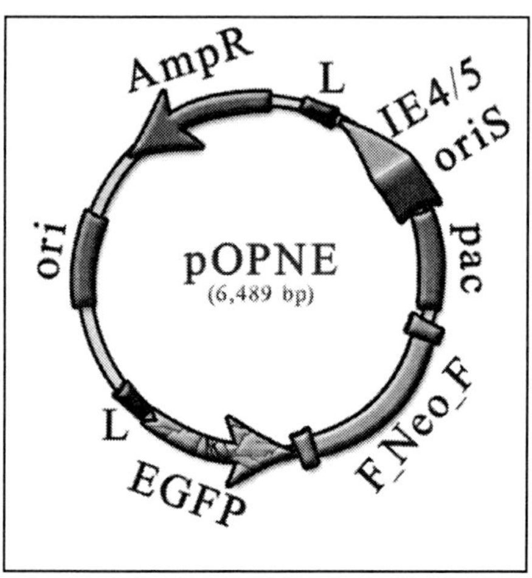

Figure 6. Structure of pOPNE amplicon plasmid. *ori*: bacterial origin of DNA replication; *AmpR*: Ampicillin resistance gene; *L*: loxP sites; *oriS*: HSV-1 origin of DNA replication; *pac*: HSV-1 cleavage/packaging-encapsidation sequence; *IE4/5*: HSV-1 IE4/5 promoter; *F_Neo_F*: Neomycin resistance gene bordered by two FRT sites (F); and *EGFP*: EGFP reporter gene, without promoter.

Preliminary results (not shown here), indicate that when the pOPNE plasmid is transfected in Cre recombinase-expressing cell lines, the amplicon plasmid recombines at the loxP sites, thereby producing two molecules: one minicircle carrying the bacterial sequences present in the plasmid backbone (residual DNA) and a second minicircle carrying the complete amplicon module devoid of bacterial sequences (Fig. 7). In addition, the Cre recombinase-mediated deletion of the bacterial sequences brought the IE4/5 promoter upstream the promoterless-EGFP open reading frame (ORF), therefore leading to EGFP transcription. Following infection of these cells with helper HSV-1, the amplicon genome was then encapsidated into HSV-1 particles. These particles expressed EGFP upon infecting target cells, therefore demonstrating that they have lost the bacterial sequences between the promoter and the EGFP ORF. These vectors are therefore appropriate for further used in gene transfer approaches. We are currently investigating the biological properties of these bacterial DNA-free amplicon vectors.

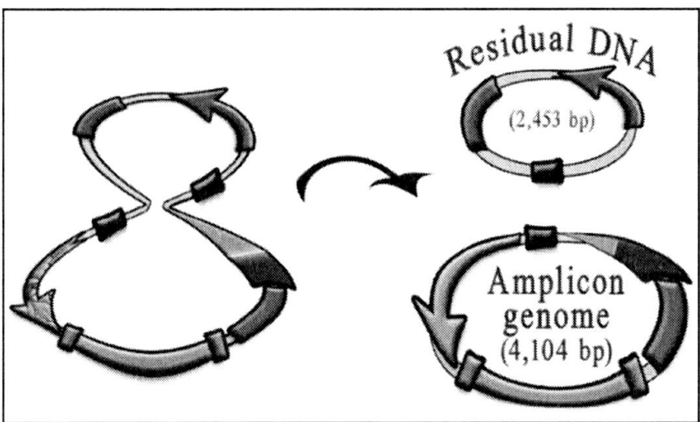

Figure 7. Cre recombinase-mediated cleavage and dissociation of pOPNE plasmid. Upon Cre recombinase expression, the amplicon plasmid pOPNE is dissociated into two DNA molecules: one carries the residual bacterial DNA and another carries the amplicon sequences devoid of bacterial DNA.

Modification of the Complementing Cell Lines

VCre4 Cell Line

As already quoted, when using HSV-1 *LaLΔJ* as helper virus, the helper particles are eliminated through a Cre/loxP-based site-specific recombination event that requires thus the expression of Cre-recombinase in the infected cells, in addition to the expression of the essential protein ICP4, whose gene is absent in this helper system. Previous cell lines expressing simultaneously ICP4 and Cre recombinase, that were used for the production of amplicon stocks with HSV-1 *LaLΔJ* helper, were derived from TE-671 cells (human rhabdomyosarcoma cells, ATCC CRL 8805) [46]. As these cells are cancer-derived cell lines, they cannot be used to produce vectors for gene therapy. In contrast, Vero and Vero-derived cell lines are approved for gene therapy approaches by the Food and Drug Administration (FDA - U.S. Department of Health and Human Services) [50].

Therefore, we have recently constructed another cell line, derived from Vero cells, which we named VCre4. These cells express Cre recombinase under the control of the HCMV promoter and the ~10.4-kbp ICP4 locus including its own promoter (HSV-1 bases 126,764-131,731). Preliminary results have shown that VCre4 cells produce higher levels of amplicon vector stocks than the previous system and these stocks contains lower than 1% of contaminant defective helper particles (data not shown here). Therefore, not only VCre4 cells provide a better packaging cell system but, in addition, the creation of this novel Vero-based cell line, brings amplicon vector technology closer to gene therapy protocols and represents a promising approach to treat neurodegenerative diseases.

Amplicon Vectors and Cancer

In this section we present a very brief summary about some past and recent applications of HSV-1 derived vectors to cancer research and experimental therapy.

HSV-1-amplicon vectors have been tested for treatment of different experimental tumours, both in brain and in other organs and tissues, exploiting its inherent absence of toxicity and ability to infect many different cell types. Since these vectors can efficiently deliver transgenes to cancer cells, but they are diluted during successive cell divisions, most studies have used acute approaches by delivering pro-drug activating enzymes, cytotoxic proteins, apoptosis-inducing factors, fusogenic proteins and small interfering RNAs (siRNAs) for growth factor receptors. In addition, several studies have targeted amplicon vector entry or expression in order to improve treatment efficacy and selectivity.

Targeting tumour cells using HSV-1 vectors is a complicated issue as the process involves multiple interactions of viral envelope glycoproteins and cellular host surface proteins. In order to improve the selectivity of amplicon vectors tropism for tumour cells, such as glioma cells, the heparan sulfate-binding domain of glycoprotein C (gC) has been replaced with a human glioma-specific peptide sequence (MG11). Preferential homing of these virions in glioma cells was confirmed in a xenograft glioma mouse model, following intravascular delivery [51]. Another important issue is the inefficient distribution of vector particles *in vivo*, which may limit their therapeutic potential in patients with gliomas and, in this context, it has been recently demonstrated that vector-mediated gene expression in gliomas was strongly dependent on vector application-injection pressure and injection time [52].

In order to improve efficiency and safety of cancer gene therapies, efforts have been made at specifically targeting proliferating cells in glioma models. The HSV-1 immediate-early gene ICP0 possesses E3-ubiquitin ligase activity [53] and it can induce the degradation of centromeric proteins [54]. Amplicons expressing the HSV-1 ICP0 were used to infect human glioblastoma Gli36 cells and well-established models of non-dividing cells, such as primary cultures of either rat cardiomyocytes or brain cells. ICP0 induced a strong cytostatic effect and significant cell death in Gli36 cells. In contrast, neither cell death nor any evidence of ICP0-induced toxicity was evident in both primary cultures of non-cycling cells. These observations suggest that ICP0 has gene therapy potential and would be the first member of a new family of cytostatic proteins that could be used to treat cancer [55].

A different approach used amplicons expressing siRNA in order to mediate post-transcriptional silencing of molecules involved in the pathogenesis of cancer or connected with the radiation resistance of tumour cells. Infected human glioblastoma cells with knockdown for the epidermal growth factor receptor (EGFR) expression displayed growth inhibition, both in culture and in athymic mice [56]. Besides, the combination of vector-mediated silencing of Rad51 expression (which is a key component of the homologous recombination repair of DNA double-strand breaks) and treatment with ionizing radiation, resulted in a pronounced reduction of human glioma cells survival in culture and in a significant decrease in tumour size in athymic mice [57].

Another tested strategy was to target tumour cells via transcriptional control of therapeutic genes. Ho and co-workers constructed a glioma-specific and cell cycle-regulated amplicon carrying the GFAP enhancer/promoter element, plus a cell cycle specific regulatory element from the cyclin A promoter [58]. Transgenic activity was mediated in a cell type-specific and cell cycle-dependent manner, both *in vitro* and *in vivo* in glioma-bearing animals [59]. Anti-tumour efficacy of this vector system was assessed using the pro-apoptotic proteins Fas ligand (FasL) and the Fas-associated death domain (FADD), inducing cell death in proliferating primary human glioma cells derived from patients, and resulting in prolonged survival of mice bearing orthotopic gliomas [60]. The authors pointed out that these vectors are stable, elicit minimal immune response and are not significantly hampered by hemotherapy or irradiation *in vivo* [58].

To achieve schwannomas regression without injury to associated neurons, it was generated an HSV-1 amplicon vector, in which the apoptosis-inducing enzyme caspase-1 (ICE), was placed under the Schwann cell-specific P0 promoter. Following direct intra-tumoral injection of the P0-ICE amplicon vector in an experimental xenograft mouse model (tumours were formed by subcutaneous injection of an immortalized human schwannoma cell line into one or both flanks of immunodeficient mice), there was a marked regression of schwannoma tumours. Injection of the same amplicon vector into the sciatic nerve produced no apparent injury to the associated dorsal root ganglia neurons or myelinated nerve fibers. Hence, the P0-ICE amplicon vector provides a potential means of 'knifeless resection' of schwannoma tumours by injection of the vector into the tumour, with low risk of damage to associated nerve fibers [61].

Amplicon Vectors and Vaccines

In this section we present a brief summary about some applications of HSV-1 derived vectors as potential vaccines.

Wild-type HSV-1 is a potent immunogen that can elicit both host innate and adaptive immune responses [62], though is also able to evade host immunity due to immuno-modulatory gene products such as the ICP47 protein [63]. The biological characteristics of HSV-1 amplicons that make them an attractive candidate for vaccine delivery applications are: its safety profile, the lack of expression of viral immuno-evasive genes, large insert capacity and ability to infect a wide range of cells, including antigen-presenting cells. So far, immunization studies involving amplicons have been essentially directed against viral pathogens and neurological disease factors.

Virus-like particles (VLPs) are promising vaccine candidates, because they represent viral antigens in the authentic conformation of the virus particles and are, therefore, readily recognized by the immune system. As VLPs do not contain genetic material, they are safer than attenuated virus vaccines. In a first study, HSV-1 amplicon vectors were constructed to coexpress the rotavirus (RV) structural genes VP2, VP6, and VP7, and were used as platforms to launch the production of RV-like particles (RVLPs) in vector-infected mammalian cells. HSV-1 amplicon vectors launching the production of heterologous rotavirus-like particles were able to induce rotavirus-specific immune response in mice [25]. In a second study, amplicons expressing Foot-and-mouth disease (FMD) virus antigens were used to generate a genetic vaccine prototype, based in the *in situ* generation of FMD-VLPs [64].

Amplicons have been also used to express antigens and elicit immune responses in the context of veterinary diseases. In particular, specific serum antibody responses were detected in mice inoculated with amplicon vectors that expressed the glycoprotein D (gD) of bovine herpesvirus [24].

Most studies focused on immunization against HIV. Hocknell and collegues showed that a single inoculation of $1x10^6$ transducing units of amplicons expressing the HIV-1 envelope glycoprotein (gp120), was able to elicit strong, antigen-specific and long-lasting ($\leq$ 5 months) cellular and humoral responses in mice [65]. Subsequent studies demonstrated that a combined heterologous prime-boost immunization approach using naked plasmid DNA prime and amplicon boost expressing the gp120 antigen could enhance by 15- to 20-fold the amplicon induced memory T-cell responses [66]. Gorantla and co-workers used non-obese diabetic/severe combined immunodeficient mice repopulated with human peripheral blood lymphocytes, as *in vivo* model for HIV immunization [67]. Injecting autologous dendritic cells transduced *ex vivo* with a gp120-expressing amplicon into these mice, resulted in primary HIV-1-specific humoral and cellular immune responses. The same authors demonstrated that there also was a partial protection against infectious HIV-1 challenge in the immunized mice [67]. Lastly, in an attempt to optimize amplicons for HIV vaccine delivery, Santos and collaborators showed that amplicons expressing the HIV-1 *gag* gene driven by two different HCMV hybrid promoters (HCMV/long-terminal repeat or HCMV/woodchuck post-regulatory element), elicited the strongest immune response in mice [68].

Other studies have explored the potential of amplicons in immunotherapy for prion disorders. Bowers and collaborators described the construction of amplicons expressing domains of PrP C fused to tetanus toxin Fragment C as adjuvant, as a base for prion disorders immunotherapy [69]. Vaccination generated significant levels of prion protein (PrP)-specific antibodies in mice, but none of the constructions was able to alter disease progression.

Since amplicons can elicit strong antigen-specific immune responses to viral or mammalian proteins, they might be used in prophylactic vaccination against pathogens or neurodegenerative diseases. However, due to the high prevalence of HSV infection within the human population, one major concern about their use as vaccines is the possible impact of pre-existing antiviral immunity on vaccine efficiency. This question remains controversial. Some studies using replication-defective HSV-1 showed strong immune responses even in the presence of anti-HSV immunity [65, 70]. In contrast, another study with a replication-defective HSV-1, showed a substantial reduction in immune responses in animals previously vaccinated against HSV-1 [71].

Amplicon Vectors and Behaviour

Amplicon vectors have been delivered into distinct brain regions to investigate complex aspects of the normal functioning of the central nervous system (CNS). Amplicon vectors designed to modify protein expression may carry sequences that allow overexpression or knockdown of an endogenous protein, or expression of an exogenous protein (e.g. dominant-negative mutants) relevant for CNS functions, like proteins directly involved in neurotransmission and in neuron signaling. A comprehensive review on the use of amplicons vectors to study behaviour can be found in Jerusalinsky and co-workers (2012) [20].

Different challenges to find causal relationships between neuronal molecular mechanisms and learning and memory processing, have been solved by the use of amplicon vectors. These vectors were used, for example, to investigate the involvement of NMDA glutamate receptor (NMDAR) in learning and memory. Amplicons carrying sequences in either sense or antisense orientations of the GluN1 subunit gene, in addition to EGFP as the reporter gene, were used to investigate the participation of hippocampal NMDAR by modifying the expression of that essential GluN1 subunit in the rat CNS. The ability to modify endogenous levels of GluN1 was first tested *in vitro* in primary cultures of neurons from rat embryo cerebral cortex [72, 73] and then assessed *in vivo.* Adult rats inoculated into the dorsal hippocampus with vectors expressing GluN1 antisense performed significantly worse than control rats in an inhibitory avoidance of a footshock task, and did not show habituation (decreased exploratory behaviour) by repeated exposure to an open field. Immunohistochemistry in serial brain slices from these animals, showed that the transduced cells represented approximately 6–7% of hippocampal pyramidal neurons in CA1 region and just about less than 1% of granule cells in the dentate gyrus, indicating that the knockdown of a single gene in a small number of those neurons significantly impaired memory [74].

Amplicon vectors expressing a constitutively active catalytic domain of the protein kinase C (PKC) β-II were used to transduce rat hippocampal dentate granule neurons. Activation of PKC pathways in a small percentage of these neurons was sufficient to enhance rat auditory discrimination reversal learning, suggesting an hippocampal auditory mediated learning in the rat [75].

Amplicon vectors have been used to elucidate the role of the alpha-amino-3-hydroxy-5-methyl-4-isoxazole-propionic acid receptor/channel (AMPAR) in learning and memory processes. Most endogenous AMPARs contain the GluR2 subunit and both inward and outward currents can pass through equally well. In contrast, AMPARs lacking GluR2 exhibit a profound inward rectification, what means that minimal outward current could pass through these channels when the cell depolarizes Thus, incorporation of recombinant AMPARs into synapses can be monitored functionally [76]. To study the AMPARs trafficking in associative fear conditioning learning, Rumpel and co-workers injected amplicon vectors expressing recombinant AMPARs subunits into the rat basolateral amygdala [77]. An amplicon vector encoding GluR1 subunit fused with GFP was used to express homomeric AMPARs, which display greater inward rectification than endogenous AMPARs, allowing electrophysiological detection of synapses undergoing plasticity by incorporation of recombinant GluR1-AMPARs. Another amplicon vector encoded the carboxyl cytoplasmic tail of GluR1 subunit fused with GFP, expressing a protein that acts as a dominant-negative construct to prevent synaptic incorporation of endogenous GluR1-receptors, and thereby blocks several forms of synaptic plasticity *in vitro* and *in vivo*. A third vector driving expression of GFP only, was used as control of infection. Using these vectors, they showed that fear conditioning drives AMPARs into the synapse of a large fraction of post-synaptic neurons in the basolateral amygdala, and that blocking GluR1-receptor trafficking in a few (~10 to 20%) neurons undergoing plasticity, was sufficient to impair memories depending on this structure [77].

The cyclic adenosine monophosphate (cAMP) response element-binding protein (CREB) plays an important role in learning and memory processes [78]. It was shown that changes in CREB function could influence the probability of individual lateral amygdala neurons to be recruited into a fear memory trace, suggesting a competitive model underlying memory formation. In such a model, eligible neurons are selected to participate in a memory trace as a function of their relative CREB activity in learning [79]. In this regard, several investigators have manipulated the function of CREB using amplicons vectors. An amplicon encoding wt-CREB was used to show that increasing CREB in the auditory thalamus enhanced memory and generalization of auditory conditioned fear, indicating that CREB-mediated plasticity in the thalamus plays a role in this cognitive process [80]. Using an amplicon vector encoding a dominant-negative mutant form of CREB (mCREB), Brightwell and collaborators demonstrated that hippocampal overexpression of mCREB can block long-term – though not short-term – memory for a socially transmitted food preference, therefore involving hippocampal CREB function in this type of memory [81]. They later showed that rats trained to make a consistent turning response in a water version of the plus maze, required

CREB function in the dorsolateral striatum to form a long-term memory of the response strategy, and showed that it is independent on CREB function in the dorsal hippocampus [82]. Using a model of protracted social isolation in adult rats, Barrot *et al* observed an increase in anxiety-like behaviour and in both the latency of the onset of sexual behaviour and the latency to ejaculate [83]. Then, in transgenic cAMP response element (CRE)-LacZ reporter mice, which express β-gal under control of CREs, they showed that protracted social isolation also reduced CRE-dependent transcription within the nucleus accumbens. This decrease in CRE-dependent transcription was mimicked in non-isolated animals by amplicon-based gene transfer of a dominant-negative mutant of CREB into the nucleus accumbens. In these animals, the local inhibition of CREB activity increased anxiety-like behaviour and delayed initiation of sexual behaviour, with no effect observed on the ejaculation parameters. In isolated animals, restoring CREB activity in the nucleus accumbens using amplicon vectors to overexpress wt-CREB, rescued the anxiety phenotype as well as the deficit in the latency to initiate sexual behaviour. This study suggests a role for the nucleus accumbens in anxiety responses and in specific aspects of sexual behaviour, and provides novel insights into the molecular mechanisms by which social interactions affect brain plasticity and behaviour [83].

Interestingly, Liu and co-workers recently reported that an amplicon expressing siRNA for the γ-Aminobutyric acid A (GABA-A) receptor α2 subunit, infused into the central nucleus of the amygdala of alcohol-preferring rats, (***i***) caused profound and selective reduction of binge drinking associated with inhibition of α2 subunit expression, (***ii***) decreased GABA-A receptor density and (***iii***) inhibited Toll-like receptor 4 (TLR4) expression [84]. Furthermore, infusion of an amplicon expressing TLR4 siRNA into the central amygdala also inhibited binge drinking, but did not cause such changes when infused into the ventral pallidum. On the other hand, binge drinking was effectively inhibited by an amplicon expressing GABA-A receptor α1 subunit siRNA, infused into the ventral pallidum, showing that TLR4 contributes to binge drinking downstream to α2 subunit in the central amygdala, but not in the ventral pallidum, underscoring the relevance of TLR4 in specific neuroanatomical sites. Those data indicate that GABA-A α2-regulated TLR4 expression in the central amygdala contributes to binge drinking and may be a key for early neuroadaptation in excessive drinking [84].

Amplicon Vectors and Neurodegenerative Diseases

Ataxias

Ataxias are a group of specific degenerative diseases of the nervous system characterized by the loss of movement coordination. Hereditary ataxias could be recessive or dominant. Nowadays, no treatment can effectively stop progression of ataxias.

The most prevalent form of hereditary ataxia is the Frederich's ataxia (FA) [85], where neurological symptoms result from neurodegeneration in the dorsal root ganglia (DRG), with loss of large sensory neurons and posterior columns, followed by degeneration in the spinocerebellar and corticospinal tracts of the spinal cord. This disease is caused by a mutation in the first intron of the frataxin gene (*frda wt*) which causes a large GAA repeat expansions (*frda mut*) and, in consequence, leads to a decrease in the level of the mitochondrial protein, frataxin.

Lim and collaborators generated a conditional frataxin transgenic mouse with the *frda wt* gene floxed (*loxP-frda*); then, they injected an amplicon vector carrying the CRE recombinase transgene into the brainstem of this transgenic *loxP-frda* mouse [86]. CRE expression causes the homologous recombination of loxP sites and the loss of *frda*, leading to a decrease in frataxin expression. These mice developed behavioral deficits after 4 weeks, resembling FA. In an attempt to generate a treatment for FA, the authors injected a second amplicon vector, bearing the cDNA of *frda* wt. These mice exhibit behavioral recovery as early as 4 weeks after the second vector injection [86].

Taking advantage of the large capacity of amplicon vectors, Gomez-Sebastian and collaborators used amplicons to deliver a 135-kb insert containing the entire *frda wt* human genomic locus, including long upstream and downstream regulatory sequences (~80-kb), to fibroblasts extracted from FA patients (which expressed low levels of frataxin) [87]. Synthesis of frataxin in these *frda wt*-transduced FA-deficient cells was confirmed by immunofluorescence [87]. The fibroblasts of FA patients have been shown to exhibit biochemical deficits, including increased sensitivity to oxidative stress. Functional complementation studies demonstrated restoration of the wild type cellular phenotype in the *frda wt*-transduced cells, in response to H_2O_2 treatment (a classical stressor) [87].

To investigate the persistence of the transgene expression, the same group injected into the adult mouse cerebellum an iBAC-HSV-1-based vector, carrying the 135-kb *frda* wt genomic DNA locus [88]. As reporter, the authors constructed another vector, but with the E. coli *lacZ* gene inserted at the ATG start codon (iBAC-*frda*-lacZ). Direct intracranial injection of this vector into the adult mouse cerebellum resulted in a large number of cells expressing lacZ; this expression was driven by the *frda wt* locus and persisted for at least 75 days. In contrast, synthesis of GFP expressed from the same vector, but driven by the HSV-1 IE4/5 promoter, was strong but transient. This study demonstrated for the first time, a sustained transgene expression *in vivo*, by amplicon delivery of a very long genomic DNA locus. All together these results suggest the potential of the HSV-1 derived-*FRDA* vectors for gene therapy of FA.

Ataxia-telangiectasia is an autosomal recessive neurodegenerative disorder due to mutations in the A-T gene (*AT*). This gene is, in the AT wt form, a kinase responsible for recognizing and correcting errors in current DNA duplication before cell division. Several mutations led to an AT-mutated: (ATM) protein with different degrees of activity, generating a pleiotropic phenotype characterized by cerebellar degeneration, immune-deficiency, cancer predisposition, increased radiation sensitivity and premature aging, according to the residual kinase activity of the expressed ATM protein [89]. As the *AT* cDNA is too large (~9-kb) for most of the currently used vectors, it was difficult to rescue the phenotype of cells expressing ATM. Therefore, Cortes and co-workers used an amplicon vector codifying the *AT* cDNA, to express a non-mutated form of AT in human fibroblasts which express ATM [90]. The AT wt protein expression was confirmed by western blot and immunofluorescence. AT kinase function was tested by *in vitro* phosphorylation of p53; the *in vivo* functionality of AT was assessed by counting the accumulation of G2/M cells after ionizing radiation [90].

In a further study, the same group constructed an amplicon encoding both the EGFP and a human FLAG-tagged-AT protein; this vector was inoculated in the cerebellum of $AT^{-/-}$ mice. The number and phenotype of infected cells were assessed by EGFP fluorescence and infection of thousands of cells at the inoculated cerebellum, including Purkinje cells, was confirmed. FLAG-tagged-AT expression was demonstrated at transcriptional (qRT-PCR, *in situ* hybridization) as well as at translational (immunoprecipitation of the full-length human protein) levels, 3 days post-inoculation [91].

In an attempt to achieve stable gene replacement, the same group generated an HSV/adeno-associated virus (AAV)-hybrid amplicon, carrying the expression cassette for the AT and EGFP cDNAs, flanked by AAV inverted terminal repeats (ITRs). In the presence of the AAV Rep proteins (proteins codified by AAV genome, required for AAV chromosome integration), this hybrid vector mediated a site-specific integration of the transgenic sequences into the AAV1 site of chromosome 19. To test functional activity of this AT vector, Cortes *et al.* exposed AT infected cells to ionizing radiation; then, AT activity was assessed by specific immunofluorescence using antibodies against the phosphorylated form of AT (pAT, activated form). An increase in pAT was observed only in AT-AAV/HSV infected cells and not in cells infected with a control vector [92]. These results showed that this AT HSV/AAV hybrid amplicon was able to integrate into the AAVS1 site and to achieve functional expression of human AT cDNA *in vivo,* in a mouse model of ATM.

Alzheimer Disease

One of the most studied neurodegenerative diseases is Alzheimer's disease (AD). In this pathology, the peptide known as amyloid beta (Aβ) acts as a neurotoxin that produces neurodegeneration. More precisely, a recently enunciated hypothesis states that soluble oligomers of Aβ peptide (named ADDLs: Aβ-derived diffusible ligands) bind to neurons, mainly at the post-synaptic side, and that this binding would be responsible for triggering toxic effects that ultimately lead to neuronal death [93, 94].

Aβ peptide is generated by degradation of the Amyloid Precursor Protein (APP). Under physiological conditions, APP is first cleaved by α-secretase and subsequently cleaved by γ-secretase, resulting in a non-amyloidogenic soluble peptide. However, under certain abnormal conditions or by blocking the normal degradation pathway, APP is cleaved by the β-secretase BACE-1 (instead of α-secretase) and then by the γ-secretase, generating amyloidogenic peptides 1-40 and 1-42, (of 40 and 42 amino acids respectively, being Aβ1-42 more amyloidogenic than Aβ1-40) [95]. Aβ initially aggregates in soluble oligomers of 2–14 monomers (ADDLs), which can bind to the post-synaptic densities; then, as the concentration rises up, they further aggregate into fibrils in the extracellular space and then form the typical amyloid plaques [94, 96-100].

Two studies have described the use of amplicons for "Aβ vaccination" in mice as a possible therapeutic strategy for AD, aimed at preventing Aβ fibrillogenesis and/or to enhance removal of parenchymal amyloid deposits. In the first study, transgenic Tg2576 mice, wich overexpress APP with the Swedish mutation that results in enhanced generation and extracellular deposition of the Aβ1–42 peptide, were injected with amplicons expressing either Aβ 1-42 (HSVAβ) or Aβ 1-42 fused to the molecular adjuvant tetanus toxin Fragment C (HSVAβ/TtxFC). Peripheral administration of both "vaccines" augmented humoral responses to Aβ and reduced CNS Aβ deposition in this model of AD. However, HSVAβ vaccination was found to be toxic, since it induced expression of pro-inflammatory transcripts within the mouse hippocampus [101].

Another amplicon vector -HSV(IE)Aβ(CMV)IL-4-, was constructed to co-delivered Aβ 1-42 and interleukin 4 (IL-4), a cytokine that promotes the generation of Th2 like T-cell responses. Triple transgenic AD (3XTg-AD) mice, which progressively develop both amyloid and neurofibrillary tangle pathology, were vaccinated with that vector or with a set of control amplicon vectors (one encoding Aβ1–42 but not IL-4, and one "empty" vector without Aβ1–42 or IL-4 transgenes). Prevention of AD-related amyloid and tau pathology progression were significantly more important in HSV(IE)Aβ (CMV)IL-4 treated mice than in control groups. Furthermore, the expression of Th2-related Aβ-specific antibodies appeared to improve learning and memory in the Barnes Maze spatial memory paradigm [102]. Therefore, these results underlined the potential of amplicons for Aβ immune-therapy of AD [103].

Neurodegenerative pathologies as AD or Parkinson's disease (PD), as well as some forms of depression, have been associated to dysfunction of receptor-neurotransmitter systems. L-glutamate is the major excitatory neurotransmitter in the CNS. Therefore, glutamate receptors represent an attractive molecular target in the treatment of neurodegenerative diseases and also in epilepsy, schizophrenia and ischemia.

There is recent evidence showing that the transmembrane protein APP appears capable of interacting with NMDAR [104, 105]. These ionotropic receptors are tetramers made of two GluN1 subunits and different GluN2 (A-D), and/or GluN3 (A-B) subunits, with GluN1 being essential for receptor assembly [106-108]. Nowadays, association of NMDAR with several neuropathologies has been continuously growing up. Thus, the generation of novel tools that modify expression and composition of NMDARs should help to understand both the normal functioning (as reported above in *"Amplicon vectors and behaviour"* section) and the physiopathology of these receptors. It was proposed that ADDLs binds to NMDAR or to post-synaptic complexes containing it, acting as gain of function ligands [93, 94, 109]. By targeting such post-synaptic complexes, ADDLs activate a cascade of signals that lead to an increase in intracellular reactive oxygen species (ROS) [93]. Decker and colleagues have demonstrated that blockade of GluN1 expression through the infection of primary cultured neurons with amplicon vectors encoding an anti-GluN1 antisense RNA, inhibited ADDLs binding to synapses [109]. In the same study, they showed that there was a great reduction in ADDL-instigated ROS formation in GluN1 knockdown neurons [109]. In addition, it has recently been reported that different regulatory GluN2 subunits would also be involved in the binding of ADDLs to synaptic sites [110, 111]. Liu and colleagues have suggested that increasing the activity of GluN2A and/or reducing that of GluN2B, may alter or reduce the expression of cytotoxic effects mediated by ADDLs in neuronal cultures [110]. On the other hand, Balducci and colleagues showed that there was an alteration in the trafficking of GluN2A and GluN2B subunits in mutant mice expressing an amyloidogenic human form of APP [111]. However, more precise studies on the possible specific interaction between different subunits of the NMDAR and the Aβ peptide, both in normal and pathological conditions, is necessary to further clarify the specific site for ADDLs binding. It should be taken into account that the decrease in GluN1, which is essential for assembly and for the membrane allocation of the receptor, would lead to a decrease of all the NMDAR subunits at the synaptic surface [112].

The genes for microtubule-associated protein tau (*MAPT*) and α-synuclein (*SNCA*) play central roles in neurodegenerative disorders. Peruzzi and colleagues have recently generated amplicon-based iBAC-vectors carrying either the 143-kb *MAPT* or the 135-kb *SNCA loci* [113]. They have used these vectors to study regulation of gene expression of both *MAPT* and *SNCA*, showing functional complementation between them in cultured neurons and in organotypic brain slices. Infected neurons were able to express functional genes under physiological regulation, including the generation of multiple splice variants. In particular, multiple *MAPT* transcripts were expressed under strict developmental and cell type-specific control. Primary cultures from *Mapt(–/–)* embryos had been shown to be resistant to Aβ peptide-induced toxicity, suggesting that the tau protein might be mediating the neurotoxicity of Aβ. To test the functionality of the *MAPT* transgene, the authors examined whether the responsiveness to Aβ peptide could be restored in the *Mapt(–/–)* neurons and in organotypic brain slices. In both preparations from *Mapt(–/–)* mice infected with the vector bearing the *MAPT* transgene, the tau protein was expressed as detected by ELISA and immunocytochemistry, and the sensitivity of *Mapt(–/–)* neurons to Aβ peptide was restored [113]. As stated by the authors, the faithful retention of gene expression and phenotypic complementation by this system provides a novel and powerful approach to analyze neurological disease genes.

Parkinson Disease

Parkinson's disease (PD) is the second most common neurodegenerative disorder. A typical feature of this disease is the progressive loss of dopaminergic neurons in the substantia nigra and a decrease in the level of dopaminergic inputs to the striatum [114]. PD is clinically characterized by akinesia or bradykinesia, rigidity and tremor that are directly related to dopaminergic striatal loss [115]. Several studies have used amplicons in experimental settings of PD. During and co-workers were the first to report the use of amplicons to deliver human tyrosine hydroxylase (TH) into the partially denervated striatum of 6-hydrodopamine-lesioned rats, used as a model of PD [116]. Efficient behavioral and biochemical recovery has persisted for one year after gene transfer.

In an attempt to improve that vector, Sun and collaborators developed another vector able to co-express two enzymes involved in the synthesis of L-DOPA: TH and the aromatic amino acid decarboxylase (AADC), under a modified neurofilament gene promoter that supports long-term expression in forebrain neurons [117]. This improved vector was injected in the right striatum of adult rats where PD-like symptoms had been previously induced by injection of 6-OHDA in the substantia nigra. Histologic analyses demonstrated neuronal-specific coexpression of TH and AADC from 4 days to 7 months after gene transfer. In these rats, vector injection was able to correct in about 80% the apomorphine-induced rotational behaviour present in this 6-OHDA rat model of PD [117]. Later on, in a series of elegant studies, Sun and co-workers further compared the activities of tissue-specific promoters to drive gene expression, particularly the promoters of TH, the neurofilament and the vesicular glutamate transport 1 (VGLUT1) [22, 118-122]. Taking advantage of the large transgene capacity of HSV derived vectors, they developed two vectors: one that co-express the three dopamine biosynthetic enzymes (TH, GTP CH I, and AADC, a 3-genes-vector) and another carrying all the three dopamine biosynthetic enzymes and the vesicular monoamine transporter (TH, GTP CH I, AADC, and VMAT-2, a 4-genes-vector). The authors compared the effects of both vectors. The 4-genes-vector supported higher levels of correction of apomorphine-induced rotational behaviour than did the 3-genes-vector, and this correction was maintained for 6 months. Proximal to the injection sites, the 4-genes-vector, but not the 3-genes-vector, supported extracellular levels of dopamine and dihydroxyphenylacetic acid (DOPAC) similar to those observed in normal rats, and only the 4-genes-vector supported significant K^+-dependent release of dopamine [118].

The effect of amplicon-mediated transduction of the dominant-negative fibroblast growth factor (FGF) receptor-1 mutant protein (FGFR1(TK-)) into the rat substantia nigra, was evaluated *in vivo* as a possible strategy to mimic the reduced FGF signaling already documented to occur in PD. Following intra-nigral delivery of the FGFR1(TK-) amplicon, the number of substantia nigra neurons expressing TH was significantly reduced, leading to the conclusion that reduced FGF signaling in the substantia nigra of Parkinsonian patients could play a role in the impaired dopaminergic transmission in PD [123]. In a further study, the same group analyzed the effect of *ex vivo* transduction of mesencephalic reaggregates with the anti-apoptotic protein bcl-2, on grafted dopamine neuron survival. Using an amplicon expressing bcl-2 under the control of the TH promoter (HSV-TH9bcl-2) to transduce mesencephalic reaggregates, it was shown that, in spite of the high efficiency of the infection (since many cells were effectively transduced), amplicon-mediated overexpression of bcl-2 did not lead to an increase in grafted TH-immune-reactive neuron number [124].

Mitochondrial alterations are detected in most neurodegenerative disorders and may contribute to the dysfunction and demise of neurons. Rotenone or 1-methyl-4-phenyl-1,2,3,6-tetrahydropyridina (MPTP) inhibit the mitochondrial complex I, causing death of dopaminergic neurons in the substantia nigra, thus providing an acute model of PD. It has been recently demonstrated that mitochondrial hexokinase II promotes neuronal survival in rotenone treated cells, and that this enzyme acts downstream of glycogen synthase kinase-3 (GSK-3), which is considered to be a critical factor in regulating neuronal cell survival and death [125]. More recently, the same group generated amplicons expressing hexokinase II; overexpression of this protein in the substantia nigra of mice subsequently administered with rotenone or MPTP, prevented neuronal cell death induced by both drugs and reduced the associated motor deficits. These results provide the first proof that hexokinase II could protect against dopaminergic neurodegeneration *in vivo,*and suggest that increase of hexokinase II expression could represent a promising approach to treat PD [126].

Narcolepsy

Narcolepsy is a neurodegenerative sleep disorder that is linked to the loss of neurons containing the neuropeptide orexin (also known as hypocretin). Liu and collaborators inoculated an amplicon vector expressing pre-pro-orexin into the lateral hypothalamus of orexin knockout mice and showed that exogenous expression of orexin significantly improved sleeping in these animals [127].

Amplicon Vectors and Nervous System Damage

Ischemia

Apoptosis plays a critical role in many neurological diseases, including stroke. To study the protective role of the antiapoptotic factor *bcl-2* in ischemia, an amplicon vector encoding bcl-2 (HSVbcl-2) was infused in rat cerebral cortex 24 hours prior to ischemia induction. Expression of bcl-2 was confirmed by immunohistochemistry in animals injected with the HSVbcl2 expression vector. Tissue viability significantly increased at the injection site in HSVbcl2, but not in animals injected with a similar vector expressing *E. coli* LacZ (HSVLacZ) [128]. Later, HSVbcl-2 was used to inject gerbils unilaterally into the CA1 region of the hippocampus, 24 hours prior to induce transient global ischemia [129]. Results showed that the local increase in bcl-2 expression using the HSVbcl-2 vector, may protect CA1 pyramidal cells from the delayed neuronal death caused by transient global ischemia, and that this happened only in the HSVbcl-2 injected hemisphere. In the same line, other group generated another bcl-2 expressing amplicon that was used to infect hippocampal cell cultures. Bcl-2 expressing vectors enhanced neuron survival after exposure to adriamycin, glutamate and hypoglycemia. Furthermore, dichlorofluorescein measurements indicated that there was a significant reduction in the accumulation of oxygen radicals associated with these insults [130]. Improved neuron survival could be attributed to the fact that the bcl-2 blocks nuclear Apoptosis-inducing factor (AIF) translocation [131]. In an experimental therapeutic approach, Lawrence and collaborators injected rats with this Bcl-2 expressing vector after ischemia induction, then leading to a reduction in neuronal loss [132]. Furthermore, they showed that there is a time window (30 minutes to 4 hours after reperfusion) where the injection was effective [132]. In accordance with these results, amplicon vectors expressing the inducible heat shock protein (hsp) 72 can also attenuate cerebral ischemic injury when introduced into the rat striatum, even during post-ischemia [133]. Furthermore, amplicons expressing hsp72 also protected neurons of CA1 hippocampal region from ischemia, and this protection would be mediated, at least in part, by increased expression of bcl-2 [134].

Neurotoxicity

Increases in cytoplasmic Ca^{2+} concentration can lead to neurotoxicity and neuronal death. The increase of Ca^{2+} could be induced by neurological trauma associated with aging and some neurological diseases. To prevent neurotoxic effects of cytoplasmic Ca^{2+} increases, an amplicon vector that expresses the calcium-binding protein calbindin D28K (calbindin D28KHSV) was used to infect neurons, both *in vitro* and *in vivo* [135, 136]. Cultured neurons infected with this vector responded to hypoglycemia and glutamatergic insults with decreased cytoplasmic [Ca^{2+}] measured by microfluorometry and increased neuron survival relative to controls [135]. Furthermore, *in vivo* injection of calbindin D28KHSV vector in the hippocampal dentate gyrus increases neuronal survival after application of the antimetabolite 3-acetylpyridine, and increases transynaptic neuronal survival in area CA3 following kainic acid neurotoxicity [136].

Reactive oxygen species (ROS) and oxidative stress damage plays an important role in neuronal death. Amplicon vectors expressing different antioxidant enzymes were used to counteract oxidative damages. The cDNA of catalase and glutathione peroxidase, two enzymes involved in hydrogen peroxide degradation, were subcloned into amplicon vectors. These vectors were shown to decrease neurotoxicity induced by different agents in primary cultures of hippocampus or cerebral cortex cells [137]. A further study using amplicons to express the antioxidant enzyme Cu–Zn–SOD, showed that these vectors were able to protect hippocampal neurons through the induction of glutathione peroxidase expression, though only in the case of neurons treated with sodium cyanide. The authors pointed out that the effect of the amplicons actually worsens the toxic effects of kainic acid, another classical ROS inducer, raising a cautionary note concerning gene therapy against oxidative damage [138]. Amplicons expressing glutamic acid decarboxylase (GAD67) were shown to protect non-differentiated cortical neurons from glutamate toxicity mediated by oxidative stress [139].

It was reported that the HIV glycoprotein of 120kDa (HIV-gp120) could be neurotoxic at certain doses and that the hsp70, hsp25 or hsp90 overexpression would protect neurons from HIV-gp120 effect. Therefore, Lim and collaborators overexpressed hsp70 with a HSV-amplicon vector in cultured hippocampal neurons; in this way, they demonstrated that hsp70 overexpression protected cultured hippocampal neurons from HIV-gp120 neurotoxicity [140].

Hipoglycemia could result in an insult for neurons. With amplicons expressing the rat brain glucose transporter (GT), it was shown that these vectors: ***(i)*** can maintain neuronal metabolism and reduce the extent of neurons loss in cultures, after a period of hypoglycemia [141]; ***(ii)*** protected cultured hippocampal, spinal cord and septal neurons against various necrotic insults, including hypoglycemia, glutamate and 3-nitropropionic acid [142]; and ***(iii)*** can enhance glucose uptake in adult rat hippocampus and in hippocampal cultures [143].

Neurotrophins

Neurotrophins are a family of growth factors that play important roles in the development and maintenance of the nervous system. The human brain-derived neurotrophic factor (BDNF) is one of the most important neurotrophins. The knowledge about the complex function of BDNF in mammalian nervous system is continuously expanding; it plays a key role as mediator of activity-induced long term potentiation (LTP) in the hippocampus, as well as in behaviour and memory [144]; BDNF has been implicated in neurodegenerative diseases [145], motor diseases [146] and fragile X syndrome [147]. This neurotrophin participates in maturation and function of mammalian auditory neurons. For this reason, amplicon vectors expressing BDNF were used to evaluate the feasibility of gene therapy of deafness. First, Geschwind and collaborators used an amplicon vector bearing a BDNF cDNA and E. coli β-galactosidase (HSVbdnflac), to evaluate the expression and biological activity in established cell lines and explant cultures, prepared from spiral ganglia of the murine ear [148]. Using two BDNF-responsive cell lines, PC12trkB and MG87trkB, they demonstrated efficient secretion of biologically active BDNF [148]. Also, the transduction of explanted spiral ganglia with HSVbdnflac, elicited a robust neuritic process outgrowth, that is comparable to the effect of exogenously added BDNF [148]. Later on, the amplicon vector expressing BDNF was used in mice to infect damaged spiral ganglion. Four weeks post-infection, stable production of BDNF was observed; and it supported the survival of auditory neurons by preventing their loss due to trophic factor deprivation-induced apoptosis [149]. In addition, the use of amplicons expressing BDNF promoted neuronal survival up to the same maximal level seen by adding exogenous BDNF, in a model of avian cochlear neuron cultures [150]. Other neurotrophins were implicated in the proper function of the auditory system. Amplicons expressing neurotrophin-3 (NT-3) were also used in murine cochlear explants model. After infection, the cochlear explants were exposed to cisplatin to induce destruction of hair cells and neurons in the auditory system. This toxicity, defined as ototoxicity, is a major dose-limiting side effect of cisplatin chemotherapy for cancer patients. Amplicon-mediated NT-3 transduction was shown to attenuate the ototoxic action of cisplatin, demonstrating the potency of NT-3 in protecting spiral ganglion neurons from degeneration [151]. Moreover, aged mouse injected in the peripheral auditory system with amplicon vectors expressing NT-3, showed significantly more spiral ganglion surviving neurons (SGNs) and lower incidence of cisplatin-induced apoptosis or necrosis, than those injected with a control vector [152]. Therefore, this approach seemed to be a promising treatment for prevention of chemical-induced hearing disorders and potentially for hearing degeneration due to normal aging.

Amplicon vectors were used to compare BDNF ability with that of the glial-derived neurotrophic factor (GDNF) to protect nigrostriatal neurons in a rat model of PD. According to this study, GDNF was significantly more effective than BDNF for both correcting behavioral deficits and protecting nigrostriatal dopaminergic neurons. The expression of both neurotrophic factors from the same vector was not more effective than expressing GDNF only [153]. In a further study addressing the effect of this trophic factor, it was shown that intracerebral administration of amplicon vectors expressing GDNF, following prior occlusion of the middle cerebral artery, displayed neuroprotection in ischemic injury. Treated animals showed reduced motor deficits and, after 1 month, there was a reduction in tissue loss, in glial fibrillary acidic protein (GFAP) and in caspase-3 immunostaining [154].

In neonatal rats, the combined delivery of NT-3 and the GluN2D subunit of the NMDA receptor, that was possible thanks to the large capacity of amplicon vectors, was used to strengthen monosynaptic connections in contused cords and to induce the appearance of weak but functional multi-synaptic connections, in double hemi-sected cords. On the other hand, the expression of either NT-3 or GluN2D alone failed to induce appearance of synaptic responses through the hemi-sected region of newborn rats [155].

More recently, the same group has treated adult rats with the following agents: (i) anti-Nogo-A antibodies to neutralize the growth-inhibitor Nogo-A; (ii) NT-3 via engineered fibroblasts, to promote neuron survival and plasticity; and (iii) the GluN2D subunit via an HSV-1 amplicon vector, to elevate NMDA receptor function by reversing its Mg^{2+} blockade, thereby enhancing synaptic plasticity and promoting the effects of NT-3 [156]. The combined treatment resulted in slightly better motor function in the absence of adverse effects (i.e., pain), suggesting that this novel combined treatment may help to improve function of the damaged spinal cord [156].

Neuropathic Pain

Neuropathic pain is a chronic form of pain that results from an injury to the nervous system. It could be due to multiple causes, like alcoholism, chemotherapy, diabetes, facial nerve problems among others. Amplicon vectors expressing proenkephalin (pHSV-hPPE-lacZ, SHPZ), were used to investigate the antinociceptive effect of leuenkephalin (a product of proekephalin proccesing). The amplicons were injected into the ventral periaqueductal grey [157] or into the cortex of rats [21]. Both studies showed that the injection of SHPZ, but not of a control vector only expressing the reporter gene lacZ, attenuated neuropathic pain and reduced induced hypersensitivity in rats [21, 157].

Acknowledgments

ALE and DAJ would like to thank the French *CNRS* and the Argentinean *CONICET* institutions for supporting the International Associated Laboratory (LIA) DEVENIR. Part of this work was supported by *Association Française contre les Myopathies* (AFM). MEM PhD studies were supported by AFM. CAB participation in this work was supported by the French Embassy in Argentine with a *"Bernardo Houssay"* postdoctoral position.

References

[1] Fatahzadeh, M. and R.A. Schwartz, Human herpes simplex labialis. *Clin Exp Dermatol,* 2007. 32(6): p. 625-30.

[2] Kelly, B.J., et al., Functional roles of the tegument proteins of herpes simplex virus type 1. *Virus Res,* 2009. 145(2): p. 173-86.

[3] Grunewald, K. and M. Cyrklaff, Structure of complex viruses and virus-infected cells by electron cryo tomography. *Curr Opin Microbiol,* 2006. 9(4): p. 437-42.

[4] Zhou, Z.H., et al., Visualization of tegument-capsid interactions and DNA in intact herpes simplex virus type 1 virions. *J Virol,* 1999. 73(4): p. 3210-8.

[5] Oh, J. and N.W. Fraser, Temporal association of the herpes simplex virus genome with histone proteins during a lytic infection. *J Virol,* 2008. 82(7): p. 3530-7.

[6] Bataille, D. and A.L. Epstein, Herpes simplex virus type 1 replication and recombination. *Biochimie,* 1995. 77(10): p. 787-95.

[7] McGeoch, D.J., et al., The complete DNA sequence of the long unique region in the genome of herpes simplex virus type 1. *J Gen Virol,* 1988. 69 (Pt 7): p. 1531-74.

[8] McGeoch, D.J., et al., Complete DNA sequence of the short repeat region in the genome of herpes simplex virus type 1. *Nucleic Acids Res,* 1986. 14(4): p. 1727-45.

[9] McGeoch, D.J., et al., Sequence determination and genetic content of the short unique region in the genome of herpes simplex virus type 1. *J Mol Biol,* 1985. 181(1): p. 1-13.

[10] de Silva, S. and W.J. Bowers, Herpes Virus Amplicon Vectors. *Viruses,* 2009. 1(3): p. 594-629.

[11] Vlazny, D.A. and N. Frenkel, Replication of herpes simplex virus DNA: localization of replication recognition signals within defective virus genomes. *Proc Natl Acad Sci U S A,* 1981. 78(2): p. 742-6.

[12] Hardwicke, M.A. and P.A. Schaffer, Cloning and characterization of herpes simplex virus type 1 oriL: comparison of replication and protein-DNA complex formation by oriL and oriS. *J Virol,* 1995. 69(3): p. 1377-88.

[13] Davison, A.J. and N.M. Wilkie, Nucleotide sequences of the joint between the L and S segments of herpes simplex virus types 1 and 2. *J Gen Virol,* 1981. 55(Pt 2): p. 315-31.

[14] Deiss, L.P., J. Chou, and N. Frenkel, Functional domains within the a sequence involved in the cleavage-packaging of herpes simplex virus DNA. *J Virol,* 1986. 59(3): p. 605-18.

[15] Nasseri, M. and E.S. Mocarski, The cleavage recognition signal is contained within sequences surrounding an a-a junction in herpes simplex virus DNA. *Virology,* 1988. 167(1): p. 25-30.

[16] McVoy, M.A., et al., Sequences within the herpesvirus-conserved pac1 and pac2 motifs are required for cleavage and packaging of the murine cytomegalovirus genome. *J Virol,* 1998. 72(1): p. 48-56.

[17] DeLuca, N.A., A.M. McCarthy, and P.A. Schaffer, Isolation and characterization of deletion mutants of herpes simplex virus type 1 in the gene encoding immediate-early regulatory protein ICP4. *J Virol,* 1985. 56(2): p. 558-70.

[18] Ozuer, A., et al., Evaluation of infection parameters in the production of replication-defective HSV-1 viral vectors. *Biotechnol Prog,* 2002. 18(3): p. 476-82.

[19] Spaete, R.R. and N. Frenkel, The herpes simplex virus amplicon: a new eucaryotic defective-virus cloning-amplifying vector. *Cell,* 1982. 30(1): p. 295-304.

[20] Jerusalinsky, D., M.V. Baez, and A.L. Epstein, Herpes simplex virus type 1-based amplicon vectors for fundamental research in neurosciences and gene therapy of neurological diseases. *J Physiol Paris,* 2012. 106(1-2): p. 2-11.

[21] Zou, W., et al., Intrathecal herpes simplex virus type 1 amplicon vector-mediated human proenkephalin reduces chronic constriction injury-induced neuropathic pain in rats. *Mol Med Rep,* 2011. 4(3): p. 529-33.

[22] Zhang, G.R. and A.I. Geller, A helper virus-free HSV-1 vector containing the vesicular glutamate transporter-1 promoter supports expression preferentially in VGLUT1-containing glutamatergic neurons. *Brain Res,* 2010. 1331: p. 12-9.

[23] Marconi, P., R. Manservigi, and A.L. Epstein, HSV-1-derived helper-independent defective vectors, replicating vectors and amplicon vectors, for the treatment of brain diseases. *Curr Opin Drug Discov Devel,* 2010. 13(2): p. 169-83.

[24] Blanc, A.M., et al., Induction of humoral responses to BHV-1 glycoprotein D expressed by HSV-1 amplicon vectors. *J Vet Sci,* 2012. 13(1): p. 59-65.

[25] Laimbacher, A.S., et al., HSV-1 amplicon vectors launch the production of heterologous rotavirus-like particles and induce rotavirus-specific immune responses in mice. *Mol Ther,* 2012. 20(9): p. 1810-20.

[26] Saydam, O., D.L. Glauser, and C. Fraefel, Construction and packaging of herpes simplex virus/adeno-associated virus (HSV/AAV) Hybrid amplicon vectors. *Cold Spring Harb Protoc,* 2012. 2012(3): p. 352-6.

[27] de Oliveira, A.P. and C. Fraefel, Herpes simplex virus type 1/adeno-associated virus hybrid vectors. *Open Virol J,* 2010. 4: p. 109-22.

[28] Mandegar, M.A., et al., Functional human artificial chromosomes are generated and stably maintained in human embryonic stem cells. *Hum Mol Genet,* 2011. 20(15): p. 2905-13.

[29] Moralli, D., et al., A novel human artificial chromosome gene expression system using herpes simplex virus type 1 vectors. *EMBO Rep,* 2006. 7(9): p. 911-8.

[30] Cuchet, D., et al., HSV-1 amplicon vectors: a promising and versatile tool for gene delivery. *Expert Opin Biol Ther,* 2007. 7(7): p. 975-95.

[31] Epstein, A.L., HSV-1-derived amplicon vectors: recent technological improvements and remaining difficulties--a review. *Mem Inst Oswaldo Cruz,* 2009. 104(3): p. 399-410.

[32] Fraefel, C., Gene delivery using helper virus-free HSV-1 amplicon vectors. *Curr Protoc Neurosci,* 2007. Chapter 4: p. Unit 4 14.

[33] Manservigi, R., R. Argnani, and P. Marconi, HSV Recombinant Vectors for Gene Therapy. *Open Virol J.* 4: p. 123-56.

[34] Kwong, A.D. and N. Frenkel, Herpes simplex virus amplicon: effect of size on replication of constructed defective genomes containing eucaryotic DNA sequences. *J Virol,* 1984. 51(3): p. 595-603.

[35] Boehmer, P.E. and I.R. Lehman, Herpes simplex virus DNA replication. *Annu Rev Biochem,* 1997. 66: p. 347-84.

[36] Cunningham, C. and A.J. Davison, A cosmid-based system for constructing mutants of herpes simplex virus type 1. *Virology,* 1993. 197(1): p. 116-24.

[37] Fraefel, C., et al., Helper virus-free transfer of herpes simplex virus type 1 plasmid vectors into neural cells. *J Virol,* 1996. 70(10): p. 7190-7.

[38] Heister, T.G., et al., Construction of HSV-1 BACs and their use for packaging of HSV-1-based amplicon vectors. *Methods Mol Biol,* 2004. 256: p. 241-56.

[39] Saeki, Y., X.O. Breakefield, and E.A. Chiocca, Improved HSV-1 amplicon packaging system using ICP27-deleted, oversized HSV-1 BAC DNA. *Methods Mol Med,* 2003. 76: p. 51-60.

[40] Saeki, Y., et al., Improved helper virus-free packaging system for HSV amplicon vectors using an ICP27-deleted, oversized HSV-1 DNA in a bacterial artificial chromosome. *Mol Ther, 2001.* 3(4): p. 591-601.

[41] Geller, A.I., A new method to propagate defective HSV-1 vectors. *Nucleic Acids Res,* 1988. 16(12): p. 5690.

[42] Geller, A.I., et al., An efficient deletion mutant packaging system for defective herpes simplex virus vectors: potential applications to human gene therapy and neuronal physiology. *Proc Natl Acad Sci U S A,* 1990. 87(22): p. 8950-4.

[43] Lim, F., et al., Generation of high-titer defective HSV-1 vectors using an IE 2 deletion mutant and quantitative study of expression in cultured cortical cells. *Biotechniques,* 1996. 20(3): p. 460-9.

[44] Pechan, P.A., et al., A novel 'piggyback' packaging system for herpes simplex virus amplicon vectors. *Hum Gene Ther,* 1996. 7(16): p. 2003-13.

[45] Zhang, X., et al., An efficient selection system for packaging herpes simplex virus amplicons. *J Gen Virol,* 1998. 79 (Pt 1): p. 125-31.

[46] Logvinoff, C. and A.L. Epstein, Genetic engineering of herpes simplex virus and vector genomes carrying loxP sites in cells expressing Cre recombinase. *Virology,* 2000. 267(1): p. 102-10.

[47] Logvinoff, C. and A.L. Epstein, Intracellular Cre-mediated deletion of the unique packaging signal carried by a herpes simplex virus type 1 recombinant and its relationship to the cleavage-packaging process. *J Virol,* 2000. 74(18): p. 8402-12.

[48] Zaupa, C., V. Revol-Guyot, and A.L. Epstein, Improved packaging system for generation of high-level noncytotoxic HSV-1 amplicon vectors using Cre-loxP site-specific recombination to delete the packaging signals of defective helper genomes. *Hum Gene Ther,* 2003. 14(11): p. 1049-63.

[49] Suzuki, M., K. Kasai, and Y. Saeki, Plasmid DNA sequences present in conventional herpes simplex virus amplicon vectors cause rapid transgene silencing by forming inactive chromatin. *J Virol,* 2006. 80(7): p. 3293-300.

[50] Content and Review of Chemistry, Manufacturing, and Control (CMC) Information for Human Gene Therapy Investigational New Drug Applications (INDs). U.S. Department of Health and Human Services / Food and Drug Administration, 2008.

[51] Ho, I.A., et al., Targeting human glioma cells using HSV-1 amplicon peptide display vector. *Gene Ther,* 2010. 17(2): p. 250-60.

[52] Kaestle, C., et al., Imaging herpes simplex virus type 1 amplicon vector-mediated gene expression in human glioma spheroids. *Mol Imaging,* 2011. 10(3): p. 197-205.

[53] Boutell, C., et al., Reciprocal activities between herpes simplex virus type 1 regulatory protein ICP0, a ubiquitin E3 ligase, and ubiquitin-specific protease USP7. *J Virol,* 2005. 79(19): p. 12342-54.

[54] Lomonte, P., K.F. Sullivan, and R.D. Everett, Degradation of nucleosome-associated centromeric histone H3-like protein CENP-A induced by herpes simplex virus type 1 protein ICP0. *J Biol Chem,* 2001. 276(8): p. 5829-35.

[55] Cuchet, D., et al., Characterization of antiproliferative and cytotoxic properties of the HSV-1 immediate-early ICPo protein. *J Gene Med,* 2005. 7(9): p. 1187-99.

[56] Saydam, O., et al., Herpes simplex virus 1 amplicon vector-mediated siRNA targeting epidermal growth factor receptor inhibits growth of human glioma cells in vivo. *Mol Ther,* 2005. 12(5): p. 803-12.

[57] Saydam, O., et al., HSV-1 amplicon-mediated post-transcriptional inhibition of Rad51 sensitizes human glioma cells to ionizing radiation. *Gene Ther, 2007*. 14(15): p. 1143-51.

[58] Ho, I.A., K.M. Hui, and P.Y. Lam, Targeting proliferating tumor cells via the transcriptional control of therapeutic genes. *Cancer Gene Ther,* 2006. 13(1): p. 44-52.

[59] Ho, I.A., K.M. Hui, and P.Y. Lam, Glioma-specific and cell cycle-regulated herpes simplex virus type 1 amplicon viral vector. *Hum Gene Ther,* 2004. 15(5): p. 495-508.

[60] Ho, I.A., W.H. Ng, and P.Y. Lam, FasL and FADD delivery by a glioma-specific and cell cycle-dependent HSV-1 amplicon virus enhanced apoptosis in primary human brain tumors. *Mol Cancer,* 2010. 9: p. 270.

[61] Prabhakar, S., et al., Imaging and therapy of experimental schwannomas using HSV amplicon vector-encoding apoptotic protein under Schwann cell promoter. *Cancer Gene Ther,* 2010. 17(4): p. 266-74.

[62] Whitley, R.J. and B. Roizman, Herpes simplex virus infections. *Lancet,* 2001. 357(9267): p. 1513-8.

[63] York, I.A., et al., A cytosolic herpes simplex virus protein inhibits antigen presentation to CD8+ T lymphocytes. *Cell,* 1994. 77(4): p. 525-35.

[64] D'Antuono, A., et al., HSV-1 amplicon vectors that direct the in situ production of foot-and-mouth disease virus antigens in mammalian cells can be used for genetic immunization. *Vaccine,* 2010. 28(46): p. 7363-72.

[65] Hocknell, P.K., et al., Expression of human immunodeficiency virus type 1 gp120 from herpes simplex virus type 1-derived amplicons results in potent, specific, and durable cellular and humoral immune responses. *J Virol,* 2002. 76(11): p. 5565-80.

[66] Wang, X., et al., Cellular immune responses to helper-free HSV-1 amplicon particles encoding HIV-1 gp120 are enhanced by DNA priming. *Vaccine,* 2003. 21(19-20): p. 2288-97.

[67] Gorantla, S., et al., Human dendritic cells transduced with herpes simplex virus amplicons encoding human immunodeficiency virus type 1 (HIV-1) gp120 elicit adaptive immune responses from human cells engrafted into NOD/SCID mice and confer partial protection against HIV-1 challenge. *J Virol,* 2005. 79(4): p. 2124-32.

[68] Santos, K., et al., Effect of promoter strength on protein expression and immunogenicity of an HSV-1 amplicon vector encoding HIV-1 Gag. *Vaccine,* 2007. 25(9): p. 1634-46.

[69] Tyler, C.M., et al., HSV amplicons: neuro applications. *Curr Gene Ther,* 2006. 6(3): p. 337-50.

[70] Brockman, M.A. and D.M. Knipe, Herpes simplex virus vectors elicit durable immune responses in the presence of preexisting host immunity. *J Virol,* 2002. 76(8): p. 3678-87.

[71] Lauterbach, H., et al., Reduced immune responses after vaccination with a recombinant herpes simplex virus type 1 vector in the presence of antiviral immunity. *J Gen Virol,* 2005. 86(Pt 9): p. 2401-10.

[72] Adrover, M.F., et al., Hippocampal infection with HSV-1-derived vectors expressing an NMDAR1 antisense modifies behavior. *Genes Brain Behav,* 2003. 2(2): p. 103-13.

[73] Cheli, V.T., et al., Gene transfer of NMDAR1 subunit sequences to the rat CNS using herpes simplex virus vectors interfered with habituation. *Cell Mol Neurobiol,* 2002. 22(3): p. 303-14.

[74] Cheli, V., et al., Knocking-down the NMDAR1 subunit in a limited amount of neurons in the rat hippocampus impairs learning. *J Neurochem,* 2006. 97 Suppl 1: p. 68-73.

[75] Neill, J.C., et al., Enhanced auditory reversal learning by genetic activation of protein kinase C in small groups of rat hippocampal neurons. *Brain Res Mol Brain Res,* 2001. 93(2): p. 127-36.

[76] Malinow, R. and R.C. Malenka, AMPA receptor trafficking and synaptic plasticity. *Annu Rev Neurosci,* 2002. 25: p. 103-26.

[77] Rumpel, S., et al., Postsynaptic receptor trafficking underlying a form of associative learning. *Science,* 2005. 308(5718): p. 83-8.

[78] Kandel, E.R., The molecular biology of memory storage: a dialogue between genes and synapses. *Science*, 2001. 294(5544): p. 1030-8.

[79] Han, J.H., et al., Neuronal competition and selection during memory formation. *Science,* 2007. 316(5823): p. 457-60.

[80] Han, J.H., et al., Increasing CREB in the auditory thalamus enhances memory and generalization of auditory conditioned fear. *Learn Mem,* 2008. 15(6): p. 443-53.

[81] Brightwell, J.J., et al., Hippocampal overexpression of mutant creb blocks long-term, but not short-term memory for a socially transmitted food preference. *Learn Mem,* 2005. 12(1): p. 12-7.

[82] Brightwell, J.J., et al., Transfection of mutant CREB in the striatum, but not the hippocampus, impairs long-term memory for response learning. *Neurobiol Learn Mem,* 2008. 89(1): p. 27-35.

[83] Barrot, M., et al., Regulation of anxiety and initiation of sexual behavior by CREB in the nucleus accumbens. *Proc Natl Acad Sci U S A,* 2005. 102(23): p. 8357-62.

[84] Liu, J., et al., Binge alcohol drinking is associated with GABAA alpha2-regulated Toll-like receptor 4 (TLR4) expression in the central amygdala. *Proc Natl Acad Sci U S A, 2011.* 108(11): p. 4465-70.

[85] Cossee, M., et al., Evolution of the Friedreich's ataxia trinucleotide repeat expansion: founder effect and premutations. *Proc Natl Acad Sci U S A,* 1997. 94(14): p. 7452-7.

[86] Lim, F., et al., Functional recovery in a Friedreich's ataxia mouse model by frataxin gene transfer using an HSV-1 amplicon vector. *Mol Ther,* 2007. 15(6): p. 1072-8.

[87] Gomez-Sebastian, S., et al., Infectious delivery and expression of a 135 kb human FRDA genomic DNA locus complements Friedreich's ataxia deficiency in human cells. *Mol Ther,* 2007. 15(2): p. 248-54.

[88] Gimenez-Cassina, A., et al., Infectious delivery and long-term persistence of transgene expression in the brain by a 135-kb iBAC-FXN genomic DNA expression vector. *Gene Ther,* 2011. 18(10): p. 1015-9.

[89] Verhagen, M.M., et al., Presence of ATM protein and residual kinase activity correlates with the phenotype in ataxia-telangiectasia: a genotype-phenotype study. *Hum Mutat,* 2012. 33(3): p. 561-71.

[90] Cortes, M.L., et al., HSV-1 amplicon vector-mediated expression of ATM cDNA and correction of the ataxia-telangiectasia cellular phenotype. *Gene Ther,* 2003. 10(16): p. 1321-7.
[91] Cortes, M.L., et al., Expression of human ATM cDNA in Atm-deficient mouse brain mediated by HSV-1 amplicon vector. *Neuroscience,* 2006. 141(3): p. 1247-56.
[92] Cortes, M.L., et al., Targeted integration of functional human ATM cDNA into genome mediated by HSV/AAV hybrid amplicon vector. *Mol Ther,* 2008. 16(1): p. 81-8.
[93] De Felice, F.G., et al., Abeta oligomers induce neuronal oxidative stress through an N-methyl-D-aspartate receptor-dependent mechanism that is blocked by the Alzheimer drug memantine. *J Biol Chem,* 2007. 282(15): p. 11590-601.
[94] Shankar, G.M., et al., Natural oligomers of the Alzheimer amyloid-beta protein induce reversible synapse loss by modulating an NMDA-type glutamate receptor-dependent signaling pathway. *J Neurosci,* 2007. 27(11): p. 2866-75.
[95] Thinakaran, G. and E.H. Koo, Amyloid precursor protein trafficking, processing, and function. *J Biol Chem, 2008.* 283(44): p. 29615-9.
[96] Shankar, G.M., et al., Amyloid-beta protein dimers isolated directly from Alzheimer's brains impair synaptic plasticity and memory. *Nat Med,* 2008. 14(8): p. 837-42.
[97] Haass, C. and D.J. Selkoe, Soluble protein oligomers in neurodegeneration: lessons from the Alzheimer's amyloid beta-peptide. *Nat Rev Mol Cell Biol,* 2007. 8(2): p. 101-12.
[98] Klein, W.L., Synaptic targeting by A beta oligomers (ADDLS) as a basis for memory loss in early Alzheimer's disease. *Alzheimers Dement,* 2006. 2(1): p. 43-55.
[99] Lesne, S., et al., A specific amyloid-beta protein assembly in the brain impairs memory. *Nature,* 2006. 440(7082): p. 352-7.
[100] Roselli, F., et al., Soluble beta-amyloid1-40 induces NMDA-dependent degradation of postsynaptic density-95 at glutamatergic synapses. *J Neurosci,* 2005. 25(48): p. 11061-70.
[101] Bowers, W.J., et al., HSV amplicon-mediated Abeta vaccination in Tg2576 mice: differential antigen-specific immune responses. *Neurobiol Aging,* 2005. 26(4): p. 393-407.
[102] Barnes, C.A., Memory deficits associated with senescence: a neurophysiological and behavioral study in the rat. *J Comp Physiol Psychol,* 1979. 93(1): p. 74-104.

[103] Frazer, M.E., et al., Reduced pathology and improved behavioral performance in Alzheimer's disease mice vaccinated with HSV amplicons expressing amyloid-beta and interleukin-4. *Mol Ther,* 2008. 16(5): p. 845-53.
[104] Cousins, S.L., et al., Amyloid precursor protein 695 associates with assembled NR2A- and NR2B-containing NMDA receptors to result in the enhancement of their cell surface delivery. *J Neurochem,* 2009. 111(6): p. 1501-13.
[105] Hoey, S.E., R.J. Williams, and M.S. Perkinton, Synaptic NMDA receptor activation stimulates alpha-secretase amyloid precursor protein processing and inhibits amyloid-beta production. *J Neurosci,* 2009. 29(14): p. 4442-60.
[106] Mori, H. and M. Mishina, Structure and function of the NMDA receptor channel. *Neuropharmacology*, 1995. 34(10): p. 1219-37.
[107] Monyer, H., et al., Developmental and regional expression in the rat brain and functional properties of four NMDA receptors. *Neuron*, 1994. 12(3): p. 529-40.
[108] Moriyoshi, K., et al., Molecular cloning and characterization of the rat NMDA receptor. *Nature,* 1991. 354(6348): p. 31-7.
[109] Decker, H., et al., N-methyl-D-aspartate receptors are required for synaptic targeting of Alzheimer's toxic amyloid-beta peptide oligomers. *J Neurochem,* 2010. 115(6): p. 1520-9.
[110] Liu, J., et al., Amyloid-beta induces caspase-dependent loss of PSD-95 and synaptophysin through NMDA receptors. *J Alzheimers Dis,* 2010. 22(2): p. 541-56.
[111] Balducci, C., et al., Cognitive deficits associated with alteration of synaptic metaplasticity precede plaque deposition in AbetaPP23 transgenic mice. *J Alzheimers Dis,* 2010. 21(4): p. 1367-81.
[112] Paoletti, P., Molecular basis of NMDA receptor functional diversity. *Eur J Neurosci,* 2011. 33(8): p. 1351-65.
[113] Peruzzi, P.P., et al., Physiological transgene regulation and functional complementation of a neurological disease gene deficiency in neurons. *Mol Ther,* 2009. 17(9): p. 1517-26.
[114] Hornykiewicz, O., Basic research on dopamine in Parkinson's disease and the discovery of the nigrostriatal dopamine pathway: the view of an eyewitness. *Neurodegener Dis,* 2008. 5(3-4): p. 114-7.
[115] Rodriguez-Oroz, M.C., et al., Initial clinical manifestations of Parkinson's disease: features and pathophysiological mechanisms. *Lancet Neurol,* 2009. 8(12): p. 1128-39.

[116] During, M.J., et al., Long-term behavioral recovery in parkinsonian rats by an HSV vector expressing tyrosine hydroxylase. *Science,* 1994. 266(5189): p. 1399-403.

[117] Sun, M., et al., Correction of a rat model of Parkinson's disease by coexpression of tyrosine hydroxylase and aromatic amino acid decarboxylase from a helper virus-free herpes simplex virus type 1 vector. *Hum Gene Ther,* 2003. 14(5): p. 415-24.

[118] Sun, M., et al., Coexpression of tyrosine hydroxylase, GTP cyclohydrolase I, aromatic amino acid decarboxylase, and vesicular monoamine transporter 2 from a helper virus-free herpes simplex virus type 1 vector supports high-level, long-term biochemical and behavioral correction of a rat model of Parkinson's disease. *Hum Gene Ther,* 2004. 15(12): p. 1177-96.

[119] Zhang, G.R., et al., The vesicular glutamate transporter-1 upstream promoter and first intron each support glutamatergic-specific expression in rat postrhinal cortex. *Brain Res,* 2011. 1377: p. 1-12.

[120] Zhang, G.R., et al., A tyrosine hydroxylase-neurofilament chimeric promoter enhances long-term expression in rat forebrain neurons from helper virus-free HSV-1 vectors. *Brain Res Mol Brain Res,* 2000. 84(1-2): p. 17-31.

[121] Cao, H., et al., Enhanced nigrostriatal neuron-specific, long-term expression by using neural-specific promoters in combination with targeted gene transfer by modified helper virus-free HSV-1 vector particles. *BMC Neurosci,* 2008. 9: p. 37.

[122] Gao, Q., et al., Isolation of an enhancer from the rat tyrosine hydroxylase promoter that supports long-term, neuronal-specific expression from a neurofilament promoter, in a helper virus-free HSV-1 vector system. *Brain Res,* 2007. 1130(1): p. 1-16.

[123] Corso, T.D., et al., Transfection of tyrosine kinase deleted FGF receptor-1 into rat brain substantia nigra reduces the number of tyrosine hydroxylase expressing neurons and decreases concentration levels of striatal dopamine. *Brain Res Mol Brain Res,* 2005. 139(2): p. 361-6.

[124] Sortwell, C.E., et al., Effects of ex vivo transduction of mesencephalic reaggregates with bcl-2 on grafted dopamine neuron survival. *Brain Res,* 2007. 1134(1): p. 33-44.

[125] Gimenez-Cassina, A., et al., Mitochondrial hexokinase II promotes neuronal survival and acts downstream of glycogen synthase kinase-3. *J Biol Chem,* 2009. 284(5): p. 3001-11.

[126] Corona, J.C., et al., Hexokinase II gene transfer protects against neurodegeneration in the rotenone and MPTP mouse models of Parkinson's disease. *J Neurosci Res,* 2010. 88(9): p. 1943-50.

[127] Liu, M., et al., Orexin (hypocretin) gene transfer diminishes narcoleptic sleep behavior in mice. *Eur J Neurosci,* 2008. 28(7): p. 1382-93.

[128] Linnik, M.D., et al., Expression of bcl-2 from a defective herpes simplex virus-1 vector limits neuronal death in focal cerebral ischemia. *Stroke,* 1995. 26(9): p. 1670-4; discussion 1675.

[129] Antonawich, F.J., H.J. Federoff, and J.N. Davis, BCL-2 transduction, using a herpes simplex virus amplicon, protects hippocampal neurons from transient global ischemia. *Exp Neurol,* 1999. 156(1): p. 130-7.

[130] Lawrence, M.S., et al., Overexpression of Bcl-2 with herpes simplex virus vectors protects CNS neurons against neurological insults in vitro and in vivo. *J Neurosci,* 1996. 16(2): p. 486-96.

[131] Zhao, H., et al., Bcl-2 transfection via herpes simplex virus blocks apoptosis-inducing factor translocation after focal ischemia in the rat. *J Cereb Blood Flow Metab,* 2004. 24(6): p. 681-92.

[132] Lawrence, M.S., et al., Herpes simplex viral vectors expressing Bcl-2 are neuroprotective when delivered after a stroke. *J Cereb Blood Flow Metab,* 1997. 17(7): p. 740-4.

[133] Hoehn, B., et al., Overexpression of HSP72 after induction of experimental stroke protects neurons from ischemic damage. *J Cereb Blood Flow Metab,* 2001. 21(11): p. 1303-9.

[134] Kelly, S., et al., Gene transfer of HSP72 protects cornu ammonis 1 region of the hippocampus neurons from global ischemia: influence of Bcl-2. *Ann Neurol,* 2002. 52(2): p. 160-7.

[135] Meier, T.J., et al., Gene transfer of calbindin D28k cDNA via herpes simplex virus amplicon vector decreases cytoplasmic calcium ion response and enhances neuronal survival following glutamatergic challenge but not following cyanide. *J Neurochem*, 1998. 71(3): p. 1013-23.

[136] Phillips, R.G., et al., Calbindin D28K gene transfer via herpes simplex virus amplicon vector decreases hippocampal damage in vivo following neurotoxic insults. *J Neurochem,* 1999. 73(3): p. 1200-5.

[137] Wang, H., et al., Over-expression of antioxidant enzymes protects cultured hippocampal and cortical neurons from necrotic insults. *J Neurochem,* 2003. 87(6): p. 1527-34.

[138] Zemlyak, I., et al., Gene therapy in the nervous system with superoxide dismutase. *Brain Res,* 2006. 1088(1): p. 12-8.

[139] Lamigeon, C., et al., Enhancement of neuronal protection from oxidative stress by glutamic acid decarboxylase delivery with a defective herpes simplex virus vector. *Exp Neurol,* 2003. 184(1): p. 381-92.

[140] Lim, M.C., S.M. Brooke, and R.M. Sapolsky, gp120 neurotoxicity fails to induce heat shock defenses, while the over expression of hsp70 protects against gp120. *Brain Res Bull,* 2003. 61(2): p. 183-8.

[141] Lawrence, M.S., et al., Herpes simplex virus vectors overexpressing the glucose transporter gene protect against seizure-induced neuron loss. *Proc Natl Acad Sci U S A,* 1995. 92(16): p. 7247-51.

[142] Ho, D.Y., et al., Defective herpes simplex virus vectors expressing the rat brain glucose transporter protect cultured neurons from necrotic insults. *J Neurochem,* 1995. 65(2): p. 842-50.

[143] Ho, D.Y., E.S. Mocarski, and R.M. Sapolsky, Altering central nervous system physiology with a defective herpes simplex virus vector expressing the glucose transporter gene. *Proc Natl Acad Sci U S A,* 1993. 90(8): p. 3655-9.

[144] Cowansage, K.K., J.E. LeDoux, and M.H. Monfils, Brain-derived neurotrophic factor: a dynamic gatekeeper of neural plasticity. *Curr Mol Pharmacol,* 2010. 3(1): p. 12-29.

[145] Gupta, V.K., et al., TrkB Receptor Signaling: Implications in Neurodegenerative, Psychiatric and Proliferative Disorders. *Int J Mol Sci,* 2013. 14(5): p. 10122-42.

[146] He, Y.Y., et al., Role of BDNF in Central Motor Structures and Motor Diseases. *Mol Neurobiol,* 2013.

[147] Castren, M. and E. Castren, BDNF in fragile X syndrome. *Neuropharmacology,* 2013.

[148] Geschwind, M.D., et al., Defective HSV-1 vector expressing BDNF in auditory ganglia elicits neurite outgrowth: model for treatment of neuron loss following cochlear degeneration. *Hum Gene Ther,* 1996. 7(2): p. 173-82.

[149] Staecker, H., et al., Brain-derived neurotrophic factor gene therapy prevents spiral ganglion degeneration after hair cell loss. *Otolaryngol Head Neck Surg,* 1998. 119(1): p. 7-13.

[150] Garrido, J.J., et al., Defining responsiveness of avian cochlear neurons to brain-derived neurotrophic factor and nerve growth factor by HSV-1-mediated gene transfer. *J Neurochem,* 1998. 70(6): p. 2336-46.

[151] Chen, X., et al., HSV amplicon-mediated neurotrophin-3 expression protects murine spiral ganglion neurons from cisplatin-induced damage. *Mol Ther,* 2001. 3(6): p. 958-63.

[152] Bowers, W.J., et al., Neurotrophin-3 transduction attenuates cisplatin spiral ganglion neuron ototoxicity in the cochlea. *Mol Ther*, 2002. 6(1): p. 12-8.

[153] Sun, M., et al., Comparison of the capability of GDNF, BDNF, or both, to protect nigrostriatal neurons in a rat model of Parkinson's disease. *Brain Res,* 2005. 1052(2): p. 119-29.

[154] Harvey, B.K., et al., HSV amplicon delivery of glial cell line-derived neurotrophic factor is neuroprotective against ischemic injury. *Exp Neurol,* 2003. 183(1): p. 47-55.

[155] Arvanian, V.L., et al., Combined delivery of neurotrophin-3 and NMDA receptors 2D subunit strengthens synaptic transmission in contused and staggered double hemisected spinal cord of neonatal rat. *Exp Neurol,* 2006. 197(2): p. 347-52.

[156] Schnell, L., et al., Combined delivery of Nogo-A antibody, neurotrophin-3 and the NMDA-NR2d subunit establishes a functional 'detour' in the hemisected spinal cord. *Eur J Neurosci,* 2011. 34(8): p. 1256-67.

[157] Zou, W., et al., Microinjection of HSV-1 amplicon vector-mediated human proenkephalin into the periaqueductal grey attenuates neuropathic pain in rats. *Int J Neurosci,* 2012. 122(4): p. 189-94.

Reviewed by

Dr. Roberto Manservigi, PhD
E-mail: mns@unife.it
Address: Università di Ferrara
Dipartimento di Scienze della Vita e Biotecnologie
Via Luigi Borsari 46
44100 – Ferrara
Italia
Phone: (+39) 0532-974470
Fax: (+39) 0532-247618

In: Advances in Viral Genomes Research ISBN: 978-1-62808-723-9
Editors: J. Borrelli and Y. Giannini

Chapter 2

Advances in Viral Genome Research of Papillomaviruses

A. C. Freitas, A. L. S. Jesus, A. P. D. Gurgel and M. A. R. Silva
Department of Genetics,
Federal University of Pernambuco, Brazil

Abstract

Papillomaviruses (PVs) infect the epithelium of amniotes, where they can cause tumours or persist asymptomatically. PVs are classified in the *Papillomaviridae* family, that contains 29 genera of PVsisolated from humans (120 types), non-human mammals, birds and reptiles (69 types). PVs have circular double-stranded DNA genomes approximately 8 kb in size and typically contain eight genes. Studies aiming the identification of PVs genomes use techniques such as PCR with consensus primers, rolling circle amplification and metagenomic methods. Advances in papillomaviral genome research have allowed the knowledge of PVs diversity and evolution of the poorly known PVs genera types, revealing that there is still a limited understanding of PVs diversity. Particularly, recent studies in Bovine Papillomavirus (BPV) have shown the identification of novel BPV types and several putative new virus types in cattle. This chapter will show new contributions in PVs genome studies.

Keywords: Papillomavirus; molecular techniques;viral metagenomics

Introduction

Papillomavirus (PVs) is a group of viruses that induce warts (or papillomas) in a variety of animals. The PVs are widespread in nature and have been recognized primarily in higher vertebrates. Viruses have been characterized from humans, cattle, rabbits, horses, dogs, sheep, elk, deer, nonhuman primates, the harvest mouse, and the multimammate mouse (*Mastomysnatalensis*) [1]. Some PVs also have malignant potential for animals and people. A number of Human Papillomaviruses (HPVs) have been implicated as the etiologic agents for cervical cancer and other epithelial tumors.

PVs are small, nonenveloped, icosahedral DNA viruses that replicate in the nucleus of squamous epithelial cells. The virions consist of a single molecule of double-stranded circular DNA about 8,000 bp in size, contained within a spherical capsid coat, composed of 72 capsomers. The genome of the PVs can be divided, in general, into three major regions: early, late, and a long control region (LCR or noncoding region [NCR])[2]. The late region of the viral genome is expressed only in differentiated layers of warts and other productive lesions, while the early region express both in warts and transformed cells [2,3]. The products of the early region encode viral regulatory proteins, including those viral proteins that are necessary for initiating viral DNA replication and transformation of the host cell [4]. The late region of PVs genomes lies downstream of the early region and encodes L1 and L2 ORFs for translation of a major (L1) and a minor (L2) capsid protein. The LCR region has no protein-coding function, but bears the origin of replication as well as multiple transcription factor binding sites that are important in regulation of RNA polymerase II-initiated transcription from viral early as well as late promoters [5].

PVs have often been classified primarily accordingly to the host species they infect and the sites or diseases with which they are associated. DNA sequencing of many PVs genomes has led their phylogenetic organization, according to the comparative homology of the L1 ORF [6]. The L1 gene is useful for classification and construction of phylogenetic trees, as it is reasonably well conserved and can be aligned for all known PVs [7].

These similarities are consistent with the conclusion that PVs have accompanied their host species during evolution and have evolved with them [8]. Although all PVs share similar genetic organization, the L1 DNA sequence identity is just above 40% between the most divergent genomes. PVs were designated as a distinct family, *Papillomaviridae*, in the 7th Report of the

ICTV [9]. De Villiers et al. [6] described the topology of phylogenetic trees based on the nucleotide sequence comparisons and biologically distinguishing features (host species, target tissues, pathogenicity, and genome organization) that determine the classification of PVs on the level of genera.

A nomenclature of these genera based on the Greek alphabet was introduced and has rapidly become accepted and widely used by the ICTV and community of PVs researchers.

Sixteen groups of PVs or individual PVs fulfilled the criterion of genera, and the Greek alphabet from the letters alpha to pi was employed to create their nomenclature. The last official classification of PVs genera ended with the genus Pi-PVs. The description of 13 new PVs genera however, exhausts the Greek alphabet. In order to create a system that continues with the Greek alphabet, it was proposed the use of the Greek alphabet a second time, employing the prefix "dyo", (i.e., Greek "a second time") [7].

Within a given genus, the L1 DNA of all members shares more than 60% identity. A viral type with a species has 71 to 89% identity with other types within the species. Within a type can exist subtypes, which share 90 to 98% identity, and variants, which have more than 98% identity [7].

Viruses can be identified by a wide range of techniques. Traditional methods include electron microscopy, cell culture, inoculation studies and serology [10].

Whereas many of the viruses known today were first identified by these techniques, the methods have limitations. For these viruses, the molecular-based techniques provide sensitive and rapid means of virus detection and identification [11].

One such approach uses sequence information from known types to identify related but undiscovered types through cross-hybridization. Another advance has involved PCR amplification of the viral genome. Such PCR-based methods comprise conventional PCR, degenerate PCR, sequence-independent PCR, and rolling circle amplification (RCA). Technological advances have also resulted in the development of metagenomics, the culture-independent study of the collective set of microbial populations in a sample by analyzing the sample's nucleotide sequence content [10].

In the next sections, it will be presented the main strategies for the identification of new types and variants of PVs and a particular broach about the recent advances in the Bovine Papillomavirus (BPV) research.

Strategies Used to Identify Papillomavirus Genomes

Since PVs life cycle requires keratinocyte differentiation, there are no conventional cell lines that allow the growth these viruses [12]. Moreover, PVs could not be transmitted to laboratory animals [13]. Hence, molecular methods have been widely used to discover and characterize the genome of new PVs as well as for diagnostics of PVs that are clinically relevant [14–37].

Among these, molecular methods such as polymerase chain reaction (PCR), hybridization, and metagenomic methodologies have been frequently used in order to detect human [14–29,31–33,35–40] and animal PVs [41–43] (Figure 1). In this section, we will discuss about several methods to identify and characterize PVs genome.

Various molecular methods, such as PCR have been used to detect specific viruses or viral families. PCR can be used to detect a few copies of a particular nucleic acid, and the viral load in a sample can be quantified with real-time PCR [44].

Although many of the PCR assays investigate the presence of a specific virus, assays have been developed to detect several viruses simultaneously. The most used molecular method to study PVs through its genome is the PCR. This methodology has been widely used to detect a broad range of PVs types in human and animals. PCR-based method has been routinely used due to high sensitivity (PCR-based method can be used to detect a few copies of PVs DNA and viral load samples) and specificity [45–48].

In addition, multiple-primed rolling circle amplification (RCA) technique, hybridization, and metagenomic methods have became a convenient tool for molecular biological research on PVs [10]. Thus, the development of molecular techniques reflected advances in knowledge of PVs genomes. Particularly, viral metagenomics have provided a powerful technology to investigate the viral flora of healthy and sick human and animals [49]. By applying these methodologies to PVs studies, it is possible to gain a better understanding in the etiology of PVs, as well as deepen the knowledge into the PVs and others virus circulating in nature, such as PVs in flies, ticks, fomites and body fluids.

Also, it may provide a better understanding of the complex interaction between virus and host.

PCR-Based Methods

Nowadays, there are several molecular assays that use PCR methods for investigate PVs. These methodologies essentially differ in the PVs gene segment amplified and in the number of primers used. For clinical studies of PVs prevalence, the most frequently used primer sets are MY9/MY11, FAP59/64, and PGMY9/ PGMY11. These consensus primer sets were designed to amplify a partial sequence of highly conserved PVs L1 gene. Also, these primers are made with the use of degenerate nucleotides that increase the range of PVs detection, however, may lower its sensitivity [50]. The MY9/MY11 primer set is the most commonly method to detect HPV infection in cervical samples [51]. These primers amplify approximately 450 base pair (bp) fragments and they allow amplification of 47 HPV DNA types. The PGMY09/PGMY11 primer set is an extended version of MY09/11 primer set, which permits to increase the sensitivity to detect HPV types [52]. Alternatively, the association of fixed nucleotide primers, such as GP5+/GP6+, with degenerate nucleotide primers may be employed to increase the sensitivity to HPV.

An alternative PCR approach using primers FAP59/64 also amplify a broad spectrum of HPV types. As well as the MY09/11 primer set, the FAP59/64 primers have degenerate nucleotides and were designed to amplify a partial sequence of highly conserved PVs L1 gene [53]. The primers FAP59/64 allow the detection of 46 HPV types. Apart from HPV, FAP59/64 and MY09/11 can be used to identify a broad range of animal PVs DNA. For instance, Ogawa et al. [22] showed that MY09/11 primers allow to detect BPV1 and BPV3. Moreover, the primers FAP59/64 can detect 13 BPV types [27,28] as well as, possible putative new BPV types [22]. Furthermore, the use of FAP59/64 primers allowed to identify PVs in chimpanzees, gorillas, macaques, monkeys, lemurs, cows and elks [54].

More recently, real-time PCR has been used to detect PVs as well as viral load assessment. This method is based on the release of fluorescence at each amplification cycle, which is directly proportional to the amount of amplicon generated. With regard to Human PVs, studies have showed that real-time PCR can be used to quantify viral load, detection and genotyping in cervical lesion or cervical cancer [55–58]. Moreover, real-time PCR has been used to detect and quantify viral load of animal PV types, such as BPV1, BPV2, BPV12 [17,59–62].

Apart from the clinical importance, the use of PCR-based methods and direct sequencing allowed to demonstrate the genetic diversity within PVs. This genetic diversity of PVs DNA can occur due to synonymous and nonsynonymous mutations [35,36,63], insertions [64] and recombination [65]. These changes can alter the structure and the function of proteins and, consequently, their biological activity [66–69]. For instance, several studies have been demonstrated that variants of HPV16, HPV18, HPV31, HPV33, HPV58 and other HPV types are related to persistent infection, progression and oncogenicity [66,67,69–71]. Moreover, studies that used PCR-based method allowed demonstrating that the genetic diversity within HPV16 and HPV18 DNA co-evolved with the three major human phylogenetic branches: African, Caucasian and Asian. Therefore, the PCR-based method is an important approach used in clinical as well as evolutionary studies with regard PVs.

Although PCR-based methods have high sensitivity and specificity, this methodology presents some disadvantages, such as the necessity of specific primer to detect PVs and short amplified fragments. In contrast, multiple primed RCA method has been used to detect new PVs. The rolling circle sequence-independent amplification technique makes use of the property of circular DNA molecules such as plasmids or viral genomes replicating through a rolling circle mechanism. Briefly, the RCA method is a DNA synthesis reaction that uses the phi29 DNA polymerase to amplify circular DNA molecules [72].

The uses of random hexamers primers, which bind at multiple locations on a circular DNA template, allow the amplification of circular DNA without requiring prior knowledge of the sequence. Recently, several studies showed the detection of new PVs using the RCA method. Although technically more demanding than other methods of sequence-independent amplification, the RCA approach has facilitated the identification of a novel PVs. For instance, the use of RCA allowed to identify 20 new HPV types in the *Betapapillomavirus* genus, 47 in *Gammapapillomavirus* genus, and 7 new HPV types in *Alphapapillomavirus* genus [13].

Similarly, RCA allowed to identify new animal PVs, such as the Canine Papillomavirus (CPV) [73,74], Feline Papillomavirus (FdPV) [75], Equine Papillomavirus (EcPV) [76], Mouse Papillomavirus (MusPV) [77], Bovine Papillomavirus (BPV) [78], *Bettongiapenicillata* Papillomavirus (BpPV) [79], *Cervuselaphus* Papillomavirus (CePV)[80] and *M. fascicularis* Papillomavirus type 1 (MfPV-1)[81].

Metagenomic Methods

As previously mentioned, several microorganisms cannot be cultivated, and therefore, some may go unnoticed and unstudied. Molecular methods widely employed such as PCR have the limitation of being used to detect specific viruses or viral families. However, metagenomics is a more general approach to study the genetic composition of uncultured samples [82]. Viral metagenomics investigates the complete genetic viral population in the sample studied [83]. These studies have also contributed to knowledge of undiscovered or low studied PVs due to its presence in no conventional sites.

In the first assays of metagenomics, the products of the amplification technologies have usually been cloned into bacteria to create libraries. Then, these cloned products were characterized by Sanger sequencing to identify any potential viruses. However, this process is quite laborious, and due to a combination of the high background of contaminating host nucleic acid and the occasionally low levels of virus, a vast number of clones may have to be sequenced before a viral sequence is identified [49].One of these techniques, the DNase-SISPA assays includes sequence-independent amplification, cloning and sequencing of amplified viral nucleic fragments followed by in silico searches for sequence similarities to known viruses. This method requires a prior digestion of non-pathogen specific nucleic acid before viral DNA/RNA. DNase-SISPA technique have been widely used for identifying known and unknown viruses [84–86], including PVs [33].

Other independent platforms were introduced for sequencing with no need for traditional cloning after amplification and yield several hundred thousand sequence reads in one run. These techniques have different sequencing principles but all have high throughput (relative to Sanger sequencing) in common and are often referred to as next-generation sequencing or high throughput sequencing [87, 88]. For metagenomic studies, 454 sequencing is often used, mainly due to the longer read length, which makes de novo assembly easier.

Next-generation sequencing (NGS)-based metagenomic have allowed the identification of several PVs, including new PVs. Hence, the use of sequencing technologies such as the 454 sequencing platform and high throughput sequencing (metagenomic sequencing and analysis) has permitted to identify new PVs in several biological samples. With regard to Human PVs, the use of metagenomic method identified the genomes of HPV116 in rectal swab [89] and nine putative HPV in cutaneous lesions skin [90]. Moreover, the use of this methodology seems to be important to indentify PVs infection in

HPV-negative samples [91] as well as sites of less common infections [92]. With regard to animal PVs, the metagenomic method allowed to identify BPV10 in teat wart of cattle [33], PVs in mosquitoes [42], a novel *Miniopterusschreibersii* Papillomavirus type 1 (MscPV1) and *Rousettusaegyptiacus* Papillomavirus type 1(RaP1) in bats [38, 93] and Tornovirus 1 (STTV1) in sea turtle [42].

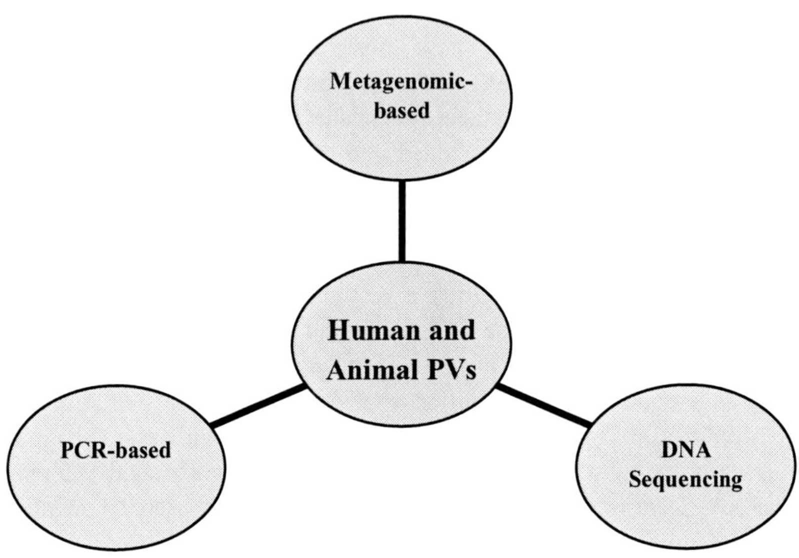

Figure 1. Current methodologies used to detect Human and Animal PVs.

A Model for Research in Papillomavirus: Advances in Bovine Papillomavirus Knowledge

Among PVs, BPV have a major role in veterinary medicine. They can cause benign and malignant lesions in cattle and are associated with horse, zebra, buffalo, yaks, giraffes, tapirs and bison lesions [17, 26, 40, 94, 95]. BPV has been widely distributed in cattle worldwide [96]. There are reports of the presence of BPV in the Americas [28], Europe [97], Asia [98], Oceania [99] and Africa [26]. BPV is considered the etiological agent of Bovine Papillomatosis and cancers of the bladder and upper gastrointestinal tract in its natural host. Although there is no global survey on the economic impact of

BPV, the numbers about world cattle population that exceeded 1 billion head of cattle reflect the importance of studying BPV. Developing countries like India, Brazil and China have the largest cattle population in the world, whereas USA, Brazil and China are the largest beef producer, respectively [100]. Cutaneous papillomas are responsible for significant economic damages to farmers due to the retarded growth of the animals, loss of weight caused by the reduction in food consumption and decrease in milk production [101,102]. Since lesions are commonly found in young animals whose immune system is still developing, its appearance is an important observation for calves because of the morbidity and mortality and high costs of treatment [103].

The presence of mucosal papillomas or tumors may progress to cancer, and the upper gastrointestinal tract and the urinary bladder are the most common locations for the development of BPV-associated carcinomas [104]. The highest economic damages caused by BPV are found in cattle from tropical and subtropical regions due to the large presence of bracken fern and its consequent chronic ingestion by the animals [104].

To date, 13 BPV types have been identified and classified into three genera: *Deltapapillomavirus* (BPV1, 2 and 13), *Epsilonpapillomavirus* (BPV5 and 8) and *Xipapillomavirus* (BPV3, 4, 6, 9, 10, 11 and 12). One BPV type is still unclassified (BPV7) [78]. All BPV types characterized so far are associated with different histopathological lesions. *Xipapillomaviruses* are pure epitheliotropic (causing true papillomas), *Deltapapillomaviruses* induce fibropapillomas and *Epsilonpapillomaviruses* can cause both types of lesions [105]. Members of *Deltapapillomavirus* genus are also associated with infections in non-epithelial sites such as blood and semen [27,106].

Besides the importance of BPV in veterinary medicine, BPV has been also largely investigated as animal model to better understand the transforming activity of PVs. BPV attracted the interest of molecular biologists since it was the first PVs able to induce transformation in cultured non-epithelial cells; furthermore its genome was the first among PVs to be completely sequenced [107]. Despite its importance, the understanding of BPV diversity is limited, probably underestimated. In contrast with HPV that presents more than 150 described HPV genomes, less than 70 non-human PVs species have been isolated and sequenced so far. However, the numbers of studies have attempted to assess the diversity and characterization of BPV genome has increased in recent years. Beyond 13 BPVs described, about 30 new putative types were isolated from various geographical regions in countries such as Brazil, Japan and Sweden. For example, BPV7, 8, 9 and 10 were designated from putative BPV types BAPV6, 2, BPV type I and BPV type II, respectively

[14, 22, 54, 98, 108–111]. These studies usually employ the PCR technique with consensus primers, and as commonly these studies describe new putative types and subtypes of BPV, it is believed that there may be an underestimation of the identification of BPV types [112].

BPV are also classified into subtypes and variants, which are useful to the knowledge of biology and diversification of PVs. The differences found in nucleotide sequences of variants of PVs could be responsible for changes in the oncogenic potential, cellular location, host immune response, function of PVs proteins (affecting pathogenesis), or binding capacity of a transcription factor [5, 35, 113, 114]. A few studies related to BPV genetic variability were done. The majority of these studies were associated with equine sarcoid, aiming at understanding differences between the disease in bovine and equine [109,115].

The improvement of molecular techniques for detection of HPV allowed the discovery of these various putative new BPV types, besides the discovery of frequent multiple infections in cattle caused by various BPV types in a single skin lesion [28, 97, 116]. BPV DNA is detected by a variety of PCR-based techniques. These PCRs are based frequently on the detection of one or two BPV types using degenerated or type-specific primers. Genotyping is performed either by real-time detection [33] or by sequence analysis [117] or restriction fragment length polymorphism (RFLP) analyses [18] of the generated PCR fragments. Usually the discovery of putative new types, subtypes and variant are made in works that employs consensus primers capable of identifying potentially more than two BPV types [22]. This methodology has suggested the existence of numerous yet uncharacterized BPV types in cattle herds from diverse geographical regions. Beyond the use of consensus primers designed for HPV, such as FAP59/FAP64 and MY09/MY11 successful in identifying cutaneous and mucosal PVs types, respectively [118,119], other consensus primers designed exclusively for L1 of BPV have been used [98].

There are reports indicating that consensus primers sometimes fail to diagnose a Papillomavirus infection because of sequence differences between consensus primers and putative PV types [54].Beyond the use of these primers, the multiple-primed RCA technique has been applied to amplify whole genomes of PVs [120,121] and also BPV [111] and became a convenient tool for molecular biological research on PVs. Also, in cattle other sequence-independent amplification techniques like DNase-SISPA have been use for characterizing genome of BPV. Due to this plasticity, BPV genomes and BPV-like genome has been detected in several sites and hosts [27, 27, 96,

122], and metagenomic methods are very suitable for samples like plasma, serum, soft tissues, respiratory secretions, cerebrospinal fluid, urine, faeces and filterable environmental [83, 86]. Therefore, these new methods seem to be very suitable for BPV studies.

The advances in knowledge of BPV genomes and putative new genomes have been reflected by the development of molecular techniques and enabled new strategies of studies with the purpose of prevention and treatment of Papillomaviruses in cattle and other animals.

References

[1] Howley PM: Papillomaviruses and Their Replication. In: Knipe, D.M., Howley, P.M. (Eds.), Fields Virology, vol. 2. Lippinicott, Williams & Wilkins, pp. 3634–3687.

[2] Zheng Z-M, Baker CC: Papillomavirus genome structure, expression, and post-transcriptional regulation. *Front Biosci. J. Virtual Libr.* 2006;11:2286–2302.

[3] Jia R, Zheng Z-M: Regulation of Bovine Papillomavirus type 1 gene expression by RNA processing. *Front Biosci. J. Virtual Libr.* 2009

[4] Campo MS. Bovine Papillomavirus: old system, new lessons? In: Campo MS (ed) Papillomavirus research: from natural history to vaccine and beyond. *Caister Academic Press*, (2006) Wymondham.

[5] Bernard H-U: Gene expression of genital Human Papillomaviruses and considerations on potential antiviral approaches. *Antivir. Ther.* 2002;7:219–237.

[6] De Villiers E-M, Fauquet C, Broker TR, Bernard H-U, zur Hausen H: Classification of Papillomaviruses. *Virology* 2004 20;324:17–27.

[7] Bernard H-U, Burk RD, Chen Z, van Doorslaer K, Hausen H zur, de Villiers E-M: Classification of Papillomaviruses (PVs) based on 189 PV types and proposal of taxonomic amendments. *Virology* 2010 25;401:70–79.

[8] Bernard H-U, Calleja-Macias IE, Dunn ST: Genome variation of Human Papillomavirus types: phylogenetic and medical implications. *Int. J. Cancer J. Int. Cance*r 2006 1;118:1071–1076.

[9] Van Regenmortel M, International, Fauquet C: Virus Taxonomy: Classification and Nomenclature of Viruses: Seventh Report of the International Committee on Taxonomy of Viruses. *Academic Press,* 2000.

[10] Bexfield N, Kellam P: Metagenomics and the molecular identification of novel viruses. *Vet. J. Lond. Engl.* 2011;190:191–198.

[11] Belak S, Thoren P, LeBlanc N, Viljoen G: Advances in viral disease diagnostic and molecular epidemiological technologies. *Expert Rev. Mol. Diagn.* 2009;9:367–381.

[12] Geimanen J, Isok-Paas H, Pipitch R, Salk K, Laos T, Orav M, et al.: Development of a cellular assay system to study the genome replication of high- and low-risk mucosal and cutaneous Human Papillomaviruses. *J. Virol.* 2011 1;85:3315–3329.

[13] De Villiers E-M: Cross-roads in the classification of Papillomaviruses. *Virology* DOI: 10.1016/j.virol.2013.04.023

[14] Maeda Y, Shibahara T, Wada Y, Kadota K, Kanno T, Uchida I, et al.: An outbreak of teat papillomatosis in cattle caused by Bovine Papillomavirus (BPV) type 6 and unclassified BPVs. *Vet. Microbiol.* 2007;121:242–248.

[15] Xi LF, Demers GW, Koutsky LA, Kiviat NB, Kuypers J, Watts DH, et al.: Analysis of Human Papillomavirus Type 16 Variants Indicates Establishment of Persistent Infection. *J. Infect. Dis.* 1995;172:747–755.

[16] Wu Y, Chen Y, Li L, Yu G, He Y, Zhang Y: Analysis of mutations in the E6/E7 oncogenes and L1 gene of Human Papillomavirus 16 cervical cancer isolates from China. *J. Gen. Virol.* 2006;87:1181 –1188.

[17] Bogaert L, Martens A, Kast WM, Van Marck E, De Cock H: Bovine Papillomavirus DNA can be detected in keratinocytes of equine sarcoid tumors. *Vet. Microbiol.* 2010 15;146:269–275.

[18] Carr EA, Théon AP, Madewell BR, Griffey SM, Hitchcock ME: Bovine Papillomavirus DNA in neoplastic and nonneoplastic tissues obtained from horses with and without sarcoids in the western United States. Am. *J. Vet. Res.* 2001;62:741–744.

[19] Stocco dos Santos RC, Lindsey CJ, Ferraz OP, Pinto JR, Mirandola RS, Benesi FJ, et al.: Bovine Papillomavirus transmission and chromosomal aberrations: an experimental model. *J. Gen. Virol.* 1998;79:2127 –2135.

[20] Wosiacki SR, Claus MP, Alfieri AF, Alfieri AA: Bovine Papillomavirus type 2 detection in the urinary bladder of cattle with chronic enzootic haematuria. *Memórias Inst. Oswaldo Cruz* 2006;101:635–638.

[21] Borzacchiello G, Ambrosio V, Roperto S, Poggiali F, Tsirimonakis E, Venuti A, et al.: Bovine Papillomavirus type 4 in oesophageal papillomas of cattle from the south of Italy. *J. Comp. Pathol.* 2003;128:203–206.

[22] Ogawa T, Tomita Y, Okada M, Shinozaki K, Kubonoya H, Kaiho I, et al.: Broad-spectrum detection of Papillomaviruses in bovine teat papillomas and healthy teat skin. *J. Gen. Virol.* 2004;85:2191 –2197.

[23] Van Dyk E, Bosman AM, van Wilpe E, Williams JH, Bengis RG, van Heerden J, et al.: Detection and characterisation of Papillomavirus in skin lesions of giraffe and sable antelope in South Africa. *J. S. Afr. Vet. Assoc.* 2011;82:80–85.

[24] Gnanamony M, Peedicayil A, Subhashini J, Ram TS, Rajasekar A, Gravitt P, et al.: Detection and quantitation of HPV 16 and 18 in plasma of Indian women with cervical cancer. *Gynecol. Oncol.* 2010;116:447–451.

[25] Ho C-M, Yang S-S, Chien T-Y, Huang S-H, Jeng C-J, Chang S-F: Detection and quantitation of Human Papillomavirus type 16, 18 and 52 DNA in the peripheral blood of cervical cancer patients. *Gynecol. Oncol.* 2005;99:615–621.

[26] Van Dyk E, Oosthuizen MC, Bosman A-M, Nel PJ, Zimmerman D, Venter EH: Detection of Bovine Papillomavirus DNA in sarcoid-affected and healthy free-roaming zebra (*Equus zebra*) populations in South Africa. *J. Virol. Methods* 2009;158:141–151.

[27] Silva MAR, Pontes NE, Da Silva KMG, Guerra MMP, Freitas AC: Detection of Bovine Papillomavirus type 2 DNA in commercial frozen semen of bulls (*Bos taurus*). *Anim. Reprod. Sci.* 2011;129:146–151.

[28] Carvalho CCR, Batista MVA, Silva MAR, Balbino VQ, Freitas AC: Detection of Bovine Papillomavirus types, co-Infection and a putative new BPV11 subtype in cattle. *Transbound. Emerg. Dis.* 2012 Jan 9; DOI: 10.1111/j.1865-1682.2011.01296.x

[29] Leishangthem GD: Detection of Bovine Papillomaviruses in cutaneous warts/papillomas in cattle. *Indian J. Vet. Pathol.* 32:15–20.

[30] Shen Z-Y, Hu S-P, Lu L-C, Tang C-Z, Kuang Z-S, Zhong S-P, et al.: Detection of Human Papillomavirus in esophageal carcinoma. *J. Med. Virol.* 2002;68:412–416.

[31] Baldez da Silva MFPT, Chagas BS, Guimaraes V, Katz LMC, Felix PM, Miranda PM, et al.: HPV31 and HPV33 incidence in cervical samples from women in Recife, *Brazil. Genet. Mol. Res.* 2009;8:1437–1443.

[32] Badaracco G, Venuti A, Morello R, Muller A, Marcante ML: Human Papillomavirus in head and neck carcinomas: prevalence, physical status and relationship with clinical/pathological parameters. *Anticancer Res.* 2000;20:1301–1305.

[33] Rai GK, Saxena M, Singh V, Somvanshi R, Sharma B: Identification of Bovine Papillomavirus 10 in teat warts of cattle by DNase-SISPA. *Vet. Microbiol.* 2011;147:416–419.

[34] Finan RR, Tamim H, Almawi WY: Identification of Chlamydia trachomatis DNA in Human Papillomavirus (HPV) positive women with normal and abnormal cytology. *Arch. Gynecol. Obstet.* 2002;266:168–171.

[35] Chagas BS, Batista MVA, Guimaraes V, Balbino VQ, Crovella S, Freitas AC: New variants of E6 and E7 oncogenes of Human Papillomavirus type 31 identified in Northeastern Brazil. *Gynecol. Oncol.* 2011;123:284–288.

[36] Chagas BS, Batista MV de A, Crovella S, Gurgel APAD, Silva Neto J da C, Serra IGSS, et al.: Novel E6 and E7 oncogenes variants of Human Papillomavirus type 31 in Brazilian women with abnormal cervical cytology. *Infect. Genet. Evol.* 2013;16C:13–18.

[37] Watts KJ, Thompson CH, Cossart YE, Rose BR: Sequence variation and physical state of Human Papillomavirus type 16 cervical cancer isolates from Australia and New Caledonia. *Int. J. Cancer J. Int. Cancer* 2002;97:868–874.

[38] Tse H, Tsang AKL, Tsoi H-W, Leung ASP, Ho C-C, Lau SKP, et al.: Identification of a novel bat Papillomavirus by metagenomics. *Plos One* 2012;7:e43986.

[39] Ge X, Li Y, Yang X, Zhang H, Zhou P, Zhang Y, et al.: Metagenomic analysis of Viruses from the bat fecal samples reveals many novel viruses in insectivorous bats in China. *J. Virol.* 2012; DOI: 10.1128/JVI.06671-11.

[40] Pangty K, Singh S, Goswami R, Saikumar G, Somvanshi R: Detection of BPV-1 and -2 and Quantification of BPV-1 by Real-Time PCR in Cutaneous Warts in Cattle and Buffaloes. *Transbound. Emerg. Dis.* 2010;57:185–196.

[41] Christian L, Claude F, Mathias A, Jessica G, Elisabeth V, Kurt T: Novel snake Papillomavirus does not cluster with other non-mammalian papillomaviruses. *Virol. J.* 2011;8:4362011.

[42] Ng TFF, Manire C, Borrowman K, Langer T, Ehrhart L, Breitbart M: Discovery of a novel single-stranded DNA virus from a sea turtle fibropapilloma by using viral metagenomics. *J. Virol.* 2009;83:2500–2509.

[43] Herbst LH, Lenz J, Van Doorslaer K, Chen Z, Stacy BA, Wellehan Jr. JFX, et al.: Genomic characterization of two novel reptilian

Papillomaviruses, *Chelonia mydas* Papillomavirus 1 and *Caretta caretta* Papillomavirus 1. *Virology* 2009;383:131–135.

[44] Mackay IM, Arden KE, Nitsche A: Real-time PCR in virology. *Nucleic. Acids Res.* 2002;30:1292–1305.

[45] Gravitt PE, Peyton C, Wheeler C, Apple R, Higuchi R, Shah KV: Reproducibility of HPV 16 and HPV 18 viral load quantitation using TaqMan real-time PCR assays. *J. Virol. Methods* 2003;112:23–33.

[46] Hesselink AT, Meijer CJLM, Poljak M, Berkhof J, Kemenade FJ van, Salm ML van der, et al.: Clinical validation of the Abbott RealTime High Risk (HR) HPV assay according to the guidelines for HPV DNA test requirements for cervical screening. *J. Clin. Microbiol.* 2013; DOI: 10.1128/JCM.00633-13

[47] Nagao S, Yoshinouchi M, Miyagi Y, Hongo A, Kodama J, Itoh S, et al.: Rapid and sensitive detection of physical status of Human Papillomavirus type 16 DNA by quantitative real-time PCR. *J. Clin. Microbiol.* 2002;40:863–867.

[48] Peitsaro P, Johansson B, Syrjänen S: Integrated Human Papillomavirus type 16 is frequently found in cervical cancer precursors as demonstrated by a novel quantitative real-time PCR technique. *J. Clin. Microbiol.* 2002;40:886–891.

[49] Blomström A-L: Applications of viral metagenomics in the veterinary field. 2010.

[50] Qu W, Jiang G, Cruz Y, Chang CJ, Ho GY, Klein RS, et al.: PCR detection of Human Papillomavirus: comparison between MY09/MY11 and GP5+/GP6+ primer systems. *J. Clin. Microbiol.* 1997;35:1304–1310.

[51] Resnick RM, Cornelissen MTE, Wright DK, Eichinger GH, Fox HS, Schegget J ter, et al.: Detection and typing of Human Papillomavirus in archival cervical cancer specimens by DNA amplification with consensus primers. *J. Natl. Cancer Inst.* 1990;82:1477–1484.

[52] Gravitt PE, Peyton CL, Alessi TQ, Wheeler CM, Coutlée F, Hildesheim A, et al.: Improved Amplification of Genital Human Papillomaviruses. *J. Clin. Microbiol.* 2000;38:357–361.

[53] Forslund O: Genetic diversity of cutaneous human papillomaviruses. *J. Gen. Virol.* 2007;88:2662–2669.

[54] Antonsson A, Hansson BG: Healthy skin of many animal species harbors Papillomaviruses which are closely related to their human counterparts. *J. Virol.* 2002;76:12537–12542.

[55] Liao Y, Zhou Y, Guo Q, Xie X, Luo E, Li J, et al.: Simultaneous detection, genotyping, and quantification of Human Papillomaviruses by multicolor real-Time PCR and melting curve analysis. *J. Clin. Microbiol.* 2013;51:429–435.

[56] Seaman WT, Andrews E, Couch M, Kojic EM, Cu-Uvin S, Palefsky J, et al.: Detection and quantitation of HPV in genital and oral tissues and fluids by real time PCR. *Virol. J.* 2010;7:194.

[57] Iftner T, Germ L, Swoyer R, Kjaer SK, Breugelmans JG, Munk C, et al.: Study comparing Human Papillomavirus (HPV) real-time multiplex PCR and hybrid capture II INNO-LiPA v2 HPV genotyping PCR Assays. *J. Clin. Microbiol.* 2009;47:2106–2113.

[58] Micalessi MI, Boulet GA, Vorsters A, De Wit K, Jannes G, Mijs W, et al.: A real-time PCR approach based on SPF10 primers and the INNO-LiPA HPV genotyping extra assay for the detection and typing of Human Papillomavirus. *J. Virol. Methods* 2013;187:166–171.

[59] Pathania S, Dhama K, Saikumar G, Shahi S, Somvanshi R: Detection and quantification of Bovine Papillomavirus type 2 (BPV-2) by real-time PCR in urine and urinary bladder lesions in enzootic bovine haematuria (EBH)-affected cows. *Transbound. Emerg. Dis.* 2012;59:79–84.

[60] Dong J, Zhu W, Goto Y, Haga T: Initial detection of a circular genome deletion in a naturally Bovine Papillomavirus-infected sample. *J. Vet. Med. Sci.* 2013;75:179–182.

[61] Haralambus R, Burgstaller J, Klukowska-Rötzler J, Steinborn R, Buchinger S, Gerber V, et al.: Intralesional Bovine Papillomavirus DNA loads reflect severity of equine sarcoid disease. *Equine Vet. J.* 2010;42:327–331.

[62] Roperto S, Russo V, Ozkul A, Sepici-Dincel A, Maiolino P, Borzacchiello G, et al.: Bovine Papillomavirus type 2 infects the urinary bladder of water buffalo (*Bubalus bubalis*) and plays a crucial role in bubaline urothelial carcinogenesis. *J. Gen. Virol.* 2012;94:403–408.

[63] De Freitas AC, Gurgel APAD, Chagas BS, Coimbra EC, do Amaral CMM: Susceptibility to cervical cancer: An overview. *Gynecol. Oncol.* 2012;126:304–311.

[64] Campitelli M, Jeannot E, Peter M, Lappartient E, Saada S, de la Rochefordiere A, et al.: Human Papillomavirus mutational insertion: specific marker of circulating tumor DNA in cervical cancer patients. *Plos One* 2012;7:e43393.

[65] Robles-Sikisaka R, Rivera R, Nollens HH, St Leger J, Durden WN, Stolen M, et al.: Evidence of recombination and positive selection in cetacean Papillomaviruses. *Virology* 2012;427:189–197.

[66] Yamada T, Manos MM, Peto J, Greer CE, Munoz N, Bosch FX, et al.: Human Papillomavirus type 16 sequence variation in cervical cancers: a worldwide perspective. *J. Virol.* 1997;71:2463–2472.

[67] Londesborough P, Ho L, Terry G, Cuzick J, Wheeler C, Singer A: Human Papillomavirus genotype as a predictor of persistence and development of high-grade lesions in women with minor cervical abnormalities. *Int. J. Cancer J. Int. Cancer* 1996;69:364–368.

[68] Zehbe I, Lichtig H, Westerback A, Lambert PF, Tommasino M, Sherman L: Rare Human Papillomavirus 16 E6 variants reveal significant oncogenic potential. *Mol. Cancer* 2011;10:77.

[69] Schiffman M, Rodriguez AC, Chen Z, Wacholder S, Herrero R, Hildesheim A, et al.: A Population-Based prospective study of carcinogenic Human Papillomavirus variant lineages, viral persistence, and cervical neoplasia. *Cancer Res*. 2010;70:3159 –3169.

[70] Cento V, Rahmatalla N, Ciccozzi M, Perno CF, Ciotti M: Intratype variations of HPV 31 and 58 in Italian women with abnormal cervical cytology. *J. Med. Virol.* 2011;83:1752–1761.

[71] Raiol T, Wyant PS, de Amorim RMS, Cerqueira DM, Milanezi N von G, Brigido M de M, et al.: Genetic variability and phylogeny of the high-risk HPV-31, -33, -35, -52, and -58 in central Brazil. *J. Med. Virol.* 2009;81:685–692.

[72] Johne R, Müller H, Rector A, van Ranst M, Stevens H: Rolling-circle amplification of viral DNA genomes using phi29 polymerase. *Trends Microbiol.* 2009;17:205–211.

[73] Lange CE, Tobler K, Lehner A, Vetsch E, Favrot C: A case of a canine pigmented plaque associated with the presence of a Xipapillomavirus. *Vet. Dermatol.* 2012;23:76–e19.

[74] Lange CE, Ackermann M, Favrot C, Tobler K: Entire Genomic Sequence of Novel Canine Papillomavirus Type 13. *J. Virol.* 2012;86:10226–10227.

[75] Lange CE, Tobler K, Markau T, Alhaidari Z, Bornand V, Stöckli R, et al.: Sequence and classification of FdPV2, a Papillomavirus isolated from feline Bowenoid in situ carcinomas. *Vet. Microbiol.* 2009;137:60–65.

[76] Scase T, Brandt S, Kainzbauer C, Sykora S, Bijmholt S, Hughes K, et al.: *Equus caballus* Papillomavirus-2 (EcPV-2): An infectious cause for equine genital cancer? *Equine Vet. J.* 2010;42:738–745.

[77] Joh J, Jenson AB, King W, Proctor M, Ingle A, Sundberg JP, et al.: Genomic analysis of the first laboratory-mouse Papillomavirus. *J. Gen. Virol.* 2011;92:692–698.

[78] Lunardi M, Alfieri AA, Otonel RAA, de Alcântara BK, Rodrigues WB, de Miranda AB, et al.: Genetic characterization of a novel Bovine Papillomavirus member of the *Deltapapillomavirus* genus. *Vet. Microbiol.* 2013;162:207–213.

[79] Bennett MD, Reiss A, Stevens H, Heylen E, Ranst MV, Wayne A, et al.: The first complete Papillomavirus genome characterized from a marsupial host: a novel isolate from *Bettongia penicillata*. *J. Virol.* 2010;84:5448–5453.

[80] Scagliarini A, Gallina L, Battilani M, Turrini F, Savini F, Lavazza A, et al.: *Cervus elaphus* Papillomavirus (CePV1): New insights on viral evolution in deer. Vet. Microbiol. DOI: 10.1016/j.vetmic.2013.03.012

[81] Joh J, Hopper K, Doorslaer KV, Sundberg JP, Jenson AB, Ghim S-J: *Macaca fascicularis* Papillomavirus type 1: a non-human primate *Betapapillomavirus* causing rapidly progressive hand and foot papillomatosis. *J. Gen. Virol.* 2009;90:987–994.

[82] Handelsman J: Metagenomics: Application of Genomics to Uncultured Microorganisms. *Microbiol. Mol. Biol. Rev.* 2004;68:669–685.

[83] Delwart EL: Viral metagenomics. Rev. Med. Virol. 2007;17:115–131.

[84] Allander T, Emerson SU, Engle RE, Purcell RH, Bukh J: A virus discovery method incorporating DNase treatment and its application to the identification of two Bovine Parvovirus species. *Proc. Natl. Acad. Sci. USA* 2001;98:11609–11614.

[85] Breitbart M, Rohwer F: Method for discovering novel DNA viruses in blood using viral particle selection and shotgun sequencing. *BioTechniques* 2005;39:729–736.

[86] Ambrose HE, Clewley JP: Virus discovery by sequence-independent genome amplification. *Rev. Med. Virol.* 2006;16:365–383.

[87] Kircher M, Kelso J: High-throughput DNA sequencing-concepts and limitations. *Bioessays* 2010;32:524–536.

[88] Shendure J, Ji H: Next-generation DNA sequencing. *Nat. Biotechnol.* 2008;26:1135–1145.

[89] Li L, Barry P, Yeh E, Glaser C, Schnurr D, Delwart E: Identification of a novel Human *Gammapapillomavirus* species. *J. Gen. Virol.* 2009;90:2413–2417.

[90] Ekström J, Bzhalava D, Svenback D, Forslund O, Dillner J: High throughput sequencing reveals diversity of Human Papillomaviruses in cutaneous lesions. *Int. J. Cancer J. Int. Cancer* 2011;129:2643–2650.

[91] Johansson H, Bzhalava D, Ekström J, Hultin E, Dillner J, Forslund O: Metagenomic sequencing of "HPV-negative" condylomas detects novel putative HPV types. *Virology* 2013;440:1–7.

[92] Mokili JL, Dutilh BE, Lim YW, Schneider BS, Taylor T, Haynes MR, et al.: Identification of a novel Human Papillomavirus by metagenomic analysis of samples from patients with febrile respiratory illness. *Plos One* 2013;8:e58404.

[93] Baker KS, Leggett RM, Bexfield NH, Alston M, Daly G, Todd S, et al.: Metagenomic study of the viruses of African straw-coloured fruit bats: Detection of a chiropteran poxvirus and isolation of a novel adenovirus. *Virology* 2013;441:95–106.

[94] Löhr CV, Juan-Sallés C, Rosas-Rosas A, García AP, Garner MM, Teifke JP: Sarcoids in Captive zebras (*Equs burcheli* II): Assocation with Bovine Papillomavirus type I infection. *J. Zoo Wildl.* Med.;36:74–81.

[95] Silvestre O, Borzacchiello G, Nava D, Iovane G, Russo V, Vecchio D, et al.: Bovine Papillomavirus Type 1 DNA and E5 Oncoprotein Expression in Water Buffalo Fibropapillomas. *Vet. Pathol. Online* 2009;46:636 – 641.

[96] Freitas AC: Recent insights into Bovine Papillomavirus. *Afr. J.* Mi*crobiol. Res.* 2011;5. DOI: 10.5897/AJMRX11.020

[97] Schmitt M, Fiedler V, Müller M: Prevalence of BPV genotypes in a German cowshed determined by a novel multiplex BPV genotyping assay. *J. Virol. Methods* 2010;170:67–72.

[98] Hatama S, Ishihara R, Ueda Y, Kanno T, Uchida I: Detection of a novel Bovine Papillomavirus type 11 (BPV-11) using *Xipapillomavirus* consensus polymerase chain reaction primers. *Arch. Virol.* 2011;156:1281–1285.

[99] Munday JS, Knight CG: Amplification of feline sarcoid-associated Papillomavirus DNA sequences from bovine skin. *Vet. Dermatol.* 2010;21:341–344.

[100] Lowe M, Gereffi G. A value chain analysis of the U.S. beef and dairy industries. *Report prepared for Environmental Defense Fund.* 2009.

[101] Meyers SA, Read WK: Squamous cell carcinoma of the vulva in a cow. *J. Am. Vet. Med. Assoc*. 1990;196:1644–1646.

[102] Scott DW, Anderson WI. Bovine cutaneous neoplasms: literature review and retrospective analysis of 62 cases (1968 to 1990). *Comp. Cont. Educ. Pract. Vet.* 1992; 14:10.

[103] Wellenberg GJ, van der Poel WHM, Van Oirschot JT: Viral infections and bovine mastitis: a review. *Vet. Microbiol.* 2002;88:27–45.

[104] Hatama S. Cutaneous papillomatosis in cattle. *J. Disaster Res*. 2012.

[105] Freitas ACF, Gurgel APAD, Chagas BS, Balbino VQ, Batista MVA. Genetic diversity of human and animal Papillomaviruses. In: Julian A, Cervantes Amaya, Miguel MFJ (Org.). *Genetic Diversity: New Research.* Hauppauge, NY: Nova Science Publishers, Inc., 2012.

[106] Silva MAR, Silva KMG, Jesus ALS, Barros LO, Corteggio A, Altamura G, et al.: The presence and gene expression of Bovine Papillomavirus in the peripheral blood and semen of healthy horses. *Transbound Emerg. Dis* 2012; DOI: 10.1111/tbed.12036

[107] Chen EY, Howley PM, Levinson AD, Seeburg PH: The primary structure and genetic organization of the Bovine Papillomavirus type 1 genome. *Nature* 1982;299:529–534.

[108] Tomita Y, Literak I, Ogawa T, Jin Z, Shirasawa H: Complete genomes and phylogenetic positions of Bovine Papillomavirus type 8 and a variant type from a European bison. *Virus Genes*. 2007;35:243–249.

[109] Claus MP, Lunardi M, Alfieri AF, Ferracin LM, Fungaro MHP, Alfieri AA: Identification of unreported putative new Bovine Papillomavirus types in Brazilian cattle herds. *Vet. Microbiol.* 2008;132:396–401.

[110] Hatama S, Nobumoto K, Kanno T: Genomic and phylogenetic analysis of two novel Bovine Papillomaviruses, BPV-9 and BPV-10. *J. Gen. Virol.* 2008;89:158 –163.

[111] Zhu W, Dong J, Shimizu E, Hatama S, Kadota K, Goto Y, et al.: Characterization of novel Bovine Papillomavirus type 12 (BPV-12) causing epithelial papilloma. *Arch. Virol*. 2012;157:85–91.

[112] Silva MAR, Batista MVA, Pontes NE, Santos EUD, Coutinho LCA, Castro RS, et al.: Comparison of two PCR strategies for the detection of Bovine Papillomavirus. *J. Virol. Methods* 2013;192:55–58.

[113] Sichero L, Villa LL: Epidemiological and functional implications of molecular variants of Human Papillomavirus. Braz. *J. Med. Biol. Res.* 2006;39:707–717.

[114] Chambers G, Ellsmore VA, O'Brien PM, Reid SWJ, Love S, Campo MS, et al.: Sequence variants of Bovine Papillomavirus E5 detected in equine sarcoids. *Virus Res.* 2003;96:141–145.

[115] Jarrett WFH, McNeil PE, Grimshaw WTR, Selman IE, McIntyre WIM: High incidence area of cattle cancer with a possible interaction between an environmental carcinogen and a papilloma virus. *Nature* 1978;274:215–217.

[116] Batista MVA, Silva MAR, Pontes NE, Reis MC, Corteggio A, Castro RS, et al.: Molecular epidemiology of bovine papillomatosis and the identification of a putative new virus type in Brazilian cattle. *Vet. J.* 2013; DOI: 10.1016/j.tvjl.2013.01.019

[117] Brandt S, Haralambus R, Schoster A, Kirnbauer R, Stanek C: Peripheral blood mononuclear cells represent a reservoir of Bovine Papillomavirus DNA in sarcoid-affected equines. *J. Gen. Virol.* 2008;89:1390 –1395.

[118] Forslund O, Antonsson A, Nordin P, Stenquist B, Hansson BG: A broad range of Human Papillomavirus types detected with a general PCR method suitable for analysis of cutaneous tumours and normal skin. *J. Gen. Virol.* 1999;80 (Pt 9):2437–2443.

[119] Manos MM, Ting Y, Wright DK, Lewis AJ, Broker TR, Wolinsky SM. The use of polymerase chain reaction amplification for the detection of genital Human Papillomaviruses. *Mol. Diagn. Human* 19897:209–214.

[120] Rector A, Bossart GD, Ghim S-J, Sundberg JP, Jenson AB, Van Ranst M: Characterization of a novel close-to-root Papillomavirus from a Florida manatee by using multiply primed rolling-circle amplification: Trichechus manatus latirostris Papillomavirus Type 1. *J. Virol.* 2004;78:12698–12702.

[121] Rector A, Tachezy R, Van Doorslaer K, MacNamara T, Burk RD, Sundberg JP, et al.: Isolation and cloning of a Papillomavirus from a North American porcupine by using multiply primed rolling-circle amplification: the *Erethizon dorsatum* Papillomavirus type 1. *Virology* 2005;331:449–456.

[122] Lindsey CL, Almeida ME, Vicari CF, Carvalho C, Yaguiu A, Freitas AC, et al.: Bovine Papillomavirus DNA in milk, blood, urine, semen, and spermatozoa of bovine papillomavirus-infected animals. *Genet. Mol. Res.* 2009;8:310–318.

In: Advances in Viral Genomes Research
Editors: J. Borrelli and Y. Giannini
ISBN: 978-1-62808-723-9

Chapter 3

Synthetic Synthesis of Viral Genomes

***Monique R. Eller*[1], *Roberto S. Dias*[2], *Rafael L. Salgado*[3], *Cynthia C. da Silva*[4] *and Sérgio O. De Paula*[2,*]**

[1]Department of Food Technology;
[2]Laboratory of Molecular Immunovirology, Department of General Biology;
[3]Department of Biochemistry and Molecular Biology;
[4]Department of Microbiology, Federal University of Viçosa, Av. PH Rolfs, s/n, Campus da UFV, Viçosa, Minas Gerais, Brazil

Abstract

Viral particles are important tools in Molecular Biology, acting as carriers of genetic material, immunogenic antigens, adjuvants, or even directly combating antibiotic-resistant microorganisms and their biofilms in hospital and industrial environments. However, the efficient use of these particles requires extensive knowledge about their characteristics and components, including those involved in their regulatory mechanisms of genome transcription and protein synthesis. The exploration of this knowledge becomes a challenge especially for scientists analyzing the

* Corresponding author: Sérgio Oliveira de Paula, E-mail address: depaula@ufv.br, Tel.: +55 31 3899-2589; fax: +55 31 3899-2549.

virions in their natural environment, due to their interactions with the complex and diverse types of biological systems, which directly influence the regulation of the infective cycles. Thus, knowledge about viral genes, their function, organization, and modulation, beyond the comprehension of the viral components as parts of complex systems, consist of the main hurdles for the controlled and predictable handling and use of these particles. For this, the technology of Synthetic Synthesis of viral genomes is distinguished from traditional genetic engineering through the use of modularity and standardization to construct proof-of-principle systems and allow generalized circuits designs to be applied to different scenarios. This new technology is made possible thanks to advances in many areas of science, from the use of restriction enzymes until the development of techniques of genomic synthesis and sequencing, like the 454 Roche, Illumina, and SOLiD systems. This technology is becoming increasingly a multidisciplinary tool used in the investigations about these complex systems, as well in the engineering of new particles for the optimization of diverse viral functions and alterations in their infectivity and affinity, or even in the development of completely new organisms and features without the need for a template. Nevertheless, advances in this technology are still limited by the lack of dynamic techniques for monitoring biological systems and efficient and standardized circuits. Here, we summarize the major characteristics of viral genomes, their organization and gene modulation, and highlight the main aspects of Synthetic Biology applied to viral genomes, as its main techniques and applications.

Introduction

Viruses are infective particles essential for the balance of ecosystems and are directly related to the evolutionary processes throughout the history of life on Earth. Therefore, the study of viral genomes attracts the interest from various areas of science, since the elucidation of the genomic sequences, its function, organization and modulation are sources of important discoveries about the evolution of species and may help to explain the biological behavior of the more complex living organisms. In addition, knowledge and handling of these genomes along with techniques of molecular biology have allowed these particles be used as vectors or gene recombination agents, as antimicrobials agents helping to control microbial contamination in industries and hospitals, or still used in the production of vaccines for the control of various pathogens, for the detection of micro-organisms, among others. For example, Lu and Collins demonstrate that an engineered bacteriophage was able to enhance the killing of antibiotic-resistant and biofilm-former bacteria, and act as a strong adjuvant for other bactericidal antibiotics (T K Lu & J J Collins 2007).

Viral genomes are extremely variable in size and structure of genes, but patterns can be observed along the evolutionary studies, especially among closely related viruses. The techniques for sequencing these genomes are already a practice and economic reality, allowing the study of genes, their function and modulation for handling these genomes can be accomplished in a rational and predictable way. Accordingly, the chemical synthesis of genes has been gradually made possible, but yet is not a reality in terms of genomes, and biological techniques are still necessary for the union of synthetically synthesized chains of nucleotides. The scientific community's interest in this issue is so meaningful that in 2012 was created a journal exclusive for this theme in which researchers could report the successful cases in the study and manipulation of viral genomes.

In this chapter we discuss about the main elements of the viral genomes, as well as modern techniques for sequencing and manipulating these genomes, concluding with an approach about the concepts of synthetic biology and examples of success in the synthesis of synthetic viral genomes reported throughout the world.

Essential Elements of Viral Genomes

Essential Genes

In the cells and viral particles, the constancy of genetic networks form biological networks, which act synergistically to perform various metabolic functions (Figure 1) (Riccione et al. 2012). Thus, a major goal of Cellular Biology is to understand how these complex networks culminate in the observed cellular behaviors. One of the most used method to assess a gene function or genetic network is based on the isolation of mutants unable to complete a particular pathway. The mutations in the viral genomes can interfere with a range of properties of these particles such as infectivity, the type of lysis plaque formed by them, their host range, or the physicochemical properties of the particles. However, the production of lethal mutations generates unviable virions, thus making them undetectable. Strategies based on synthetic biology to study genetic circuits or viral motifs gene can be a solution to this problem. Thus, to obtain more complex circuitry is necessary to know which genes confer essential characteristics to each organism, which has led to a large number of studies devoted to this subject (Little 2010).

Viruses have high genetic diversity, ranging from extremely simple genomes, such as that of porcine circovirus, which has a circular genome of only 1.7 nucleotides and three Open Reading Frames (ORFs) (Figure 2), till extremely large genomes, such that of mimivirus, which genome codes for more than 900 ORFs (Abrescia et al. 2012). There is not a universally conserved gene among all viruses. However, some genes that encode proteins required for replication and particle formation are present in all members in a group (Dolja & Koonin 2011). Although there is similarity between sequences of the viral genomes analyzed until now, surprisingly the vast majority of sequences of viruses present in ocean has low similarity to known viral sequences, finding greater similarity to bacterial genes, some involved in the main processes of the cellular metabolism. Two not mutually exclusive hypotheses could explain this issue, one based in the contamination of samples with bacterial DNA, raising the possibility of fails in that protocols of metagenomic analysis, and the other in the fact that our current knowledge on viral genomes is not representative of the current virome which makes oceans the main viromes on the planet (Kristensen et al. 2010).

Due to the great diversity of viral genomes and the distinct genetic needs between different viruses, there are no genes that are essential to all of them. However, two proteins are considered the minimum requirements for the existence of a viral particle - the viral capsid protein and - the viral polymerase. Although the genes encoding these proteins are highly variable among different groups, its existence is almost unanimous among known viral genomes. The other genes present in the viral genomes are components of transcription modules that vary between different virus groups and depend on the particle structure and their mechanism of infection/replication. Next are cited groups of some viruses and their basic replication machinery.

1. Bacteriophages as well as plasmids, can be considered genetic elements, motile or not, present in prokaryotic cells. Once internalized, the genomes of these viruses are called prophages or replicons, and replicate intracellularly by the "replicon model" using key regulatory elements that interact with an origin of replication (Ori). The replication of linear viral genomes composed of dsDNA seems to be simple, despite its unidirectional mechanism of initiation, which leads to the loss of a portion of the genome in each replication cycle. Evolutionary studies showed that phages found at least four

different mechanisms to address this issue: 1 - The *Bacillus subtilis* phage Φ29 proteins uses specific proteins as portable primers, which remain covalently attached to the ends of the viral genome; 2 - The *Escherichia coli* phage T7 possess direct repeats at the terminal ends of its genome, which are regenerated by the processing of its concatamers by enzymes called terminases during the packing of monomeric genomes, 3 – The *Escherichia coli* phage N15 regenerates the ends of its linear double-stranded DNA by the action of the specialized enzyme protelomerase, similarly to what happens for eukaryotic genomes by the enzyme telomerase, 4 - linear genomes of some phages are circularized after its injection into the host, while others are integrated into the host chromosome in the form of prophage.

2. The porcine circovirus (PCV) are small, icosahedral viruses with approximately 17 nm in diameter and composed of a circular ssDNA genome. Replication of the genome of PCV-1 is made by rolling circle (RCR) mechanism, which is widely used by phages, plasmids, animal viruses and phytoviruses (Cheung 2012). The PCV have two opposing ORFs from the Ori region. The ORF1 is transcribed and processed, giving rise to the Rep proteins (Rep and Rep ') related to viral replication. ORF2 is present in the complementary strand and is responsible for the synthesis of the viral capsid protein (Cheung 2012). The ORF3 was recently characterized as a non-structural protein, not essential for replication in cell cultures, and with a major role in the induction of apoptosis by activating the initiation of the via of caspase 8 and caspase 3in the host (Liu et al. 2005).

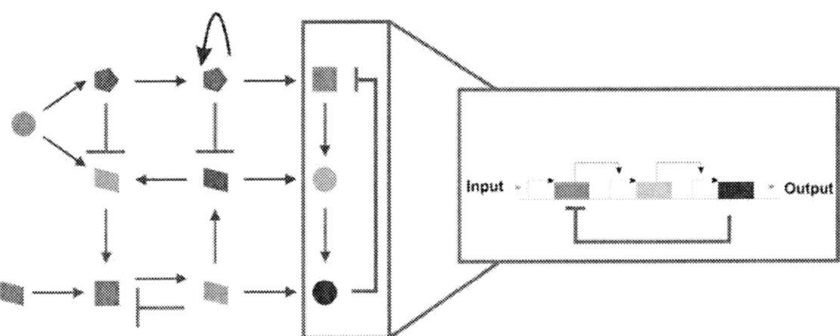

Figure 1. The more complex circuits have various integrated routes. Synthetic biology allows reconstructing parts of these circuits with different purposes, such as the study of these genes, the production of metabolites and control of cell populations (adapted from Riccione et al. 2012).

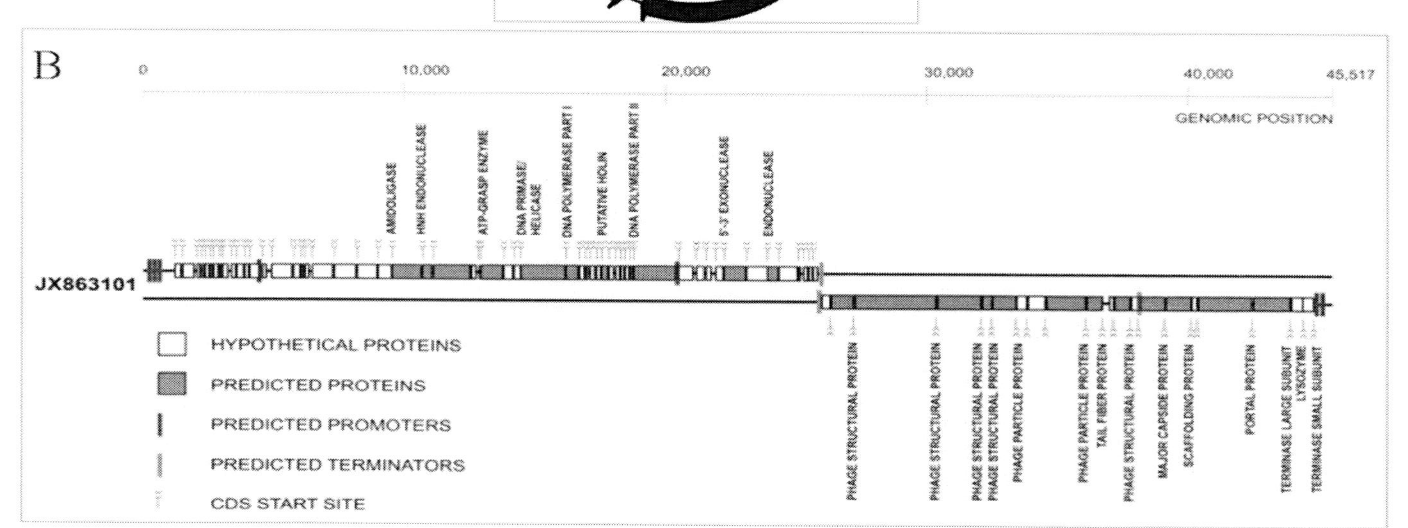

Figure 2. (Continued).

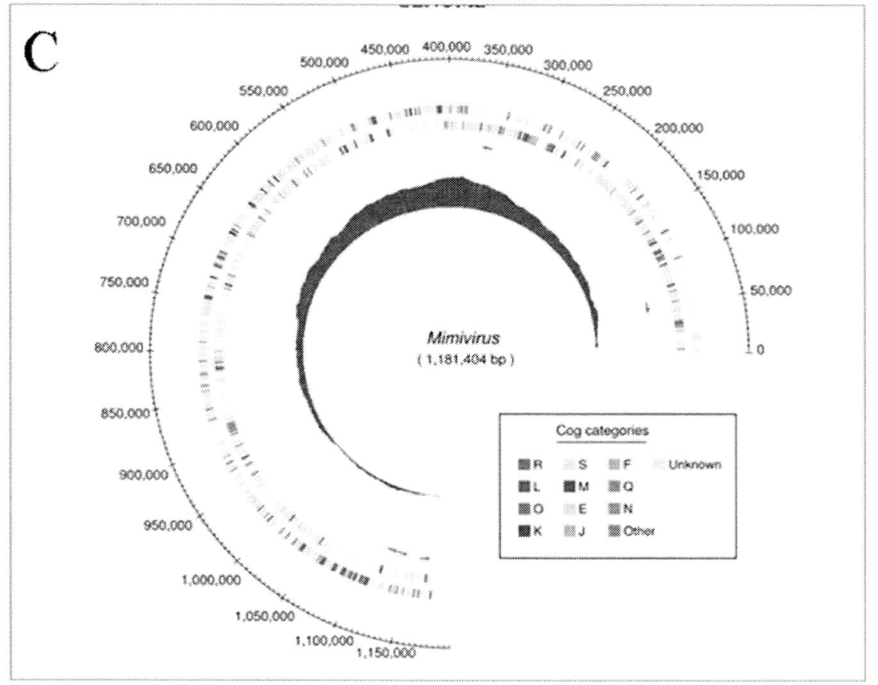

Figure 2. Genomes with different degrees of complexity. A) PCV genome, the simplest viral genome known, composed of essential genes for replication (ORF1), virion assembly (ORF2) and host apoptosis (ORF3). B) UFV-P2 genome, with intermediate complexity and at least two transcription modules of replication and virion assembly and release. C) mimivirus genome presenting high complexity, possessing 1.2 Mb and more than 900 genes (Raoult et al. 2004).

3. Viruses of the *Flaviviridae* family have linear genomes consisting of single stranded RNA (ssRNA) with positive polarity, what means that it can be transcribed immediately after its entry into the host cell. These genomes are transcribed as a single polyprotein, which is cleaved generating the viral structural and non-structural proteins. Among these is the NS5 protein, a RNA-dependent RNA polymerase (RdRp), which is essential for the phage genome replication, since natural animal cells do not possess this enzyme. A second element essential to the replication of these viruses is the sequences for cyclization, present in the terminal ends of their genomes. Recent studies show that the RdRp binds to the clamp formed after genome circularization and initiates its replication by synthesizing the negative strand RNA, which is the replicative intermediate of flaviviruses (Malet et al. 2008).
4. Until the discovery of mimiviruses, the concept of viruses was based on the filtration of solutions in membranes with 500 nm pores. However, mimivirus have a diameter of about 750 nm, beyond fibrils, which made this definition improper. Although little is known about the different stages of mimiviruses infection , it is known that their 981 genes characterize them as the first known viruses with complex genomes similar to other intracellular bacterial parasites. In addition, these viruses were the first in which components of the system of protein synthesis were observed, such as amino-acyl tRNA synthetases (Claverie & Abergel 2010). In addition to size and complexity of its genome, the mimivirus are characterized by a high gene density (90.5%), with something like 1262 ORFs, and of these 298 have function assigned to several central functions of cellular metabolism, ranging from the amino acid metabolism and transport, translation and post-translational modifications.

Organization and Modulation of Viral Genomes

The evolution of the viral particles led to the mobile interchangeable elements, known as modules, which carry specific biological functions. Thereby, viruses found ways to adapt their genomes to different niches through the combination of different modules. A detailed comparison of two genomes from phages belonging to different niches may shows similarities in genetic organization, although containing alternating regions of high and low sequence similarity (Botstein 1980). The high degree of conservation in the ordering of functional genes into transcriptional modules is evident, allowing the inference of functional proteins by comparing the gene sequences using bioinformatics (Veesler & Cambillau 2011). Furthermore, the homology in the organization of modules in phages from different families enables the establishment of hybrids between them.

The genetic modules have two basic characteristics: the efficient conducting of their biological functions and their ability to permute among genomes. Both features define the frequency at which each module is inserted (or lost) in a genome, and this decision is mainly related to the functional compatibility of a module with a variety of combinations with the other essential modules present in the genome of an organism. Thus, the presence/absence of specific viral modules in the genomes may be also indicative of the phylogenetic distance of viral hosts. Lawrence and coworkers proposed that the prophages and phage genomes could be described as an assembly of several genetic modules, which tend to remain associated independently of the recombinant nature of their genomes (Lawrence et al. 2002). Therefore, these modules can been classified according to the functions of their inner genes, as occurs in the modules of adsorption, replication, regulation, recombination, capsid, basal plate, proteins forming contractile tails, proteins forming long non-contractile tails, tail fibers, lysis, integration, excision and pathogenesis (Toussaint et al. 2007; Lima-Mendez et al. 2011).

The classification of viruses can be performed by the analysis of evolutionarily conserved modules (ECM). Some ECMs appear to be specific to certain types of phages, host groups, or type of infective cycle. For example, the ECM17 is composed of two sub-modules, the first of integration /regulation, which are real markers for integrative phages, and the replicative module found in Gram-positive-infecting phages (Pellegrini et al. 1999; Lima-Mendez et al. 2011). Recently, a software was developed based on this type of classification and was named Phage Classification Tool Set - PHACTS, which uses genomic analysis to predict the infective cycle of different phages (McNair et al. 2012).

Methodologies for Genomic Manipulation

The term Synthetic Biology was first described in 1910, by Leduc, S. (Stéphane Leduc 1910). This technology has enabled the development of cells and biological devices with features molded according to a number of different interests. This has led to a demand for more dynamic, simple, and cost effective new technologies of DNA synthesis and sequencing (T K Lu & J J Collins 2007).

Genomic Sequencing

The first regulatory circuits were discovered over 40 years and consisted of a feedback inhibitors of amino acidic biosynthetic pathways (Westerhoff & Palsson 2004). The discovery, production and commercial viability of restriction enzymes, and the standardization of techniques as molecular cloning constituted major advances in the 70s, allowing the emergence of technologies such as Genetic Engineering and Biotechnology. Already in the 1980s some of the fundamental experimental approaches of Molecular Biology were developed and improved, and in the mid 90's the first automated DNA sequencers were glimpsed, leading the genetic sequencing to the genomic scale. The automation, miniaturization and multiplexing of multiple trials led to the generation of additional types of "omic" data, such as metabolomics, proteomics, and genomics (Westerhoff & Palsson 2004; Hunkapiller et al. 1991; Rowen et al. 1997; Uetz et al. 2000).

The first methodology for sequencing DNA was the chemical method of *Sanger*, in which labeled modified nucleotides are used for reading the template DNA during amplification. The Human Genome Project was performed by a variation of this method, the *shotgun sequencing*, in which the DNA is mechanically or enzymatically broken and the fragments generated are cloned and individually sequenced. After obtaining individual sequences, these are then overlapped for the assembly of the entire genome. The new sequencing methodologies known as next generation sequencing (NGS) are based on this principle, performing random reads throughout the genome and assembling the fragments generated by overlapping (J. Zhang et al. 2011). However, the number of bases that can be continuously read by NGS methods is very small when compared to Sanger, what is the major limitation of these methods since they generate a very large amount of data, making their processing difficult and time-consuming. Some of these new platforms are described below:

1. Roche GS-FLX 454 – It was as the first commercial sequencing platform, introduced in 2004, and has as main advantage the generation of the larger fragments among the NGS platforms. The key

process is based on mixing the reactants for amplification in an emulsion. Inside the micelles, the amplification is performed in DNA-binding beads. In this platform, the DNA is read through pyrosequencing, in which light is emitted and recorded when each nucleotide is incorporated by the polymerase during the fragment amplification.

2. Illumina - It was the second platform to be commercialized and is currently the most used due to its lower costs. This platform is also based on the sequencing by synthesis, in which DNA fragments are attached to fixed supports via adapters (oligo-primed). These oligo-adapters are also used to amplify the free end of the DNA fragment, thus generating clusters of identical fragments. The amplifications using terminators nucleotides labeled generate signals that are interpreted and give rise to the sequence of each fragment.
3. SOLiD - This platform is based on sequencing by ligation, in which the templates are attached to microspheres in emulsions and combined with universal sequencing primer, the enzyme ligase, and fluorescent probes. Sequencing occurs by hybridization of such fluorescent probes with the target fragment through several steps using different combinations of universal primers.

In the last two years new genetic sequencing platforms were presented and called third generation sequencing. One of them is based on changes in electric current generated by the passage of DNA in nanopores, with an ability to read million bases per hour. A second method is based on the variation of electrical conductance of a solution, which occurs due to the polymerase addition of nucleotides to the chain of nucleic acids, presenting processivity about 22 nucleotides per second (Y. S. Chen et al. 2013; Radford et al. 2012; Eisenstein 2012).

Genetic Manipulations

The mutations in the phage genomes may be obtained from the use of physical agents such as ultraviolet light, chemical mutagens, or by recombinant DNA technology. These techniques allow changing specifically or randomly the sequences of nucleotides that compose the viral genomes, thus generating particles with different characteristics in relation to the wild ones. For example, the technique of *DNA shuffling* allows the random recombination of fragments between homologous genes, thereby generating a series of particles with different versions of a particular gene. On the other hand, *error prone PCR* is a technique of inducing errors during amplification of a specific gene, generating directed punctual mutations. Homologous recombination has also been used for deleting genes or obtaining recombinant phages. The technology known as *recombineering* allows the introduction of *in vivo* changes, such as knock outs, deletions, insertions, or point mutations, in a genome. Yu and coworkers reported the development of a transformation system of *Escherichia coli* genome by inserting the recombination system of the phage λ (Exo and Beta) in the recombination system *RecBCD* (Marinelli et al. 2012; Yu et al. 2000).

The chemical synthesis of nucleotide sequences led to a race of biotechnology companies for more efficient ways to synthesize genetic material with increasing amounts of nucleotides, making it possible to sell affordable gene sequences. Currently, it is possible to a researcher to purchase gene sequences as large as 10,000 base pairs with a delivery time of about a month! The high-throughput DNA syntesis technology provides to researchers a new and important tool. However, the big question still lies in correctly performing the biological question to be answered by this tool, which can be facilitated by detailed study and mathematical modeling of the known genetic circuits. The chemical modification of nucleotides and the generation and study of isolated circuits offer a way to study independently the influence of circuits in natural pathways without disturbing the other components of the metabolism.

Some of these circuits have been used in useful tasks, such as control of cell populations, decision making for biosensors, genetic timer for fermentation processes, and image processing. More recently these circuits have been used to solve medical and industrial solutions such as for the development of bacteria capable of invading cancer cells, for the dispersion of bacterial biofilms by engineered phages, as well as for the production of precursors of anti-malarial drugs for the generation of synthetic microbial routes. However, in such cases, engineered organisms have a single modified gene circuit, which does not include the potential offered by synthetic biology, so showing a discrepancy between the simplicity of these genetic circuits and the promise of assembly of these circuits in complex genetic networks (T K Lu et al. 2009). Although there are no limitations on the number of new circuits to be built, the number of interoperable and well-characterized parts hampers the development of more complex biological systems, thus creating a constant need to expand the set of available tools. This limits our ability to build truly modular circuits and highlights the need for an accurate characterization of the interactions between the different components of these systems so that negative interactions are minimized (T K Lu et al. 2009).

Synthesis of New Viral Genomes

The beginning of the study and synthesis of genetic circuits evolved the manipulation of the genetic material of most organisms by a series of restriction enzymes, recombination and amplification reactions, and the expression of heterologous proteins in foreign systems. While these techniques are now used routinely in laboratories around the world, the need for templates for genetic manipulation and the high complexity of the regulatory system of the host led to the need to synthesize *de novo* genomes, from which it would be possible to predict the interactions between different genetic components. Thus, the chemical synthesis of long polymer chains of nucleotides and the construction and maintenance of DNA libraries have favored the emergence of this new way of creating and manipulating entirely new biological systems.

The synthesis and manipulation of genes for their insertion or deletion in the genomes is already a strategy routinely used in different fields of science, for example, in elucidating metabolic pathways, in discovering gene function, or to genetic improvement. However, the synthesis of new genomes without templates has been done only since the last decade, and the increasing knowledge on the functions of regulatory sequences have contributed to the synthesis of larger and more complex genomes, with increasingly predictable and rational designs. An example was the synthesis *de novo* of a bacterial genome in 2010 (Gibson et al. 2010). While the synthesis of genes considered natural and proximal regulatory aspects, the genomic synthesis should consider the different regulatory elements contained in the entire genome, the interactions between them, and the products of the different genes. This is only possible from a consolidated knowledge about the viral genomes, its function and modulation.

Genetic Circuits

Synthetic biology integrates the techniques of molecular biology to the principles of engineering, computer sciences, and mathematical models. From this integration, it is possible to design and construct genetic circuits that enable living cells to perform new functions. The term genetic circuitry has been widely used to refer to well-defined sequences of genetic material that, together, perform specific functions when transcribed. This concept is directly related to the rational construction of genomes with predefined characteristics. A large number of genetic circuits have been developed, most of them switches, oscillators, digital logic gates, filters, modular and interoperable memory devices, counters, sensors and protein scaffolds (Timothy K Lu 2010). By using these circuits is possible to predict, at least partially, the features and biological modulations of synthetic viral particles, which can be manipulated to present characteristics desirable in relation to the functions for which they are used. In the future, researchers envision that, by analogy to electronic circuits, it will be possible to generate viral particles programmable from the generic combination of genetic circuits which, when combined, generate an predictable phenotype (Ferber 2004). For this, it is necessary to develop well-defined stages of construction, monitoring and debugging and setting of these circuits.

Monitoring the Expression of Genetic Elements

Practices methodologies of large-scale monitoring of the expression of circuits are essential for the development of efficient networks and for obtaining reliable and reproducible results. These techniques are mainly based on the detection of reporter elements - molecules whose concentration is proportional to the expression of the target gene, as mRNAs and proteins. there are many methods based on this concept, but the optimal methodology for monitoring gene circuits, although it is dependent on the circuit involved, may generate strong signals with low noise, present specificity, low cost, non-invasiveness, multiplexability, and, if possible, generate dynamic real-time signals (Timothy K Lu 2010). It is possible to monitor a circuit for analyzing the total protein content of an organism without the need for a reporter element. However, this method is nonspecific and often non- accepted for the quantitative determination of expression, and thus does not significantly contribute to the understanding of these circuits. Thus, some macromolecules are frequently used as reference detectors agents. Initial studies relied on the determination of the promoter activity by the simple fusion of the promoter plus the gene of interest with a particular reporter gene. These systems are very effective and used in a number of systems, although the scientists still search for alternatives that are less invasive, cheaper and applicable to a larger number of systems.

Colorimetric Assays

The colorimetric assays for monitoring protein expression consist of simple but inaccurate techniques. These assays are based mainly on the co-expression of the gene of interest with an auxiliary gene encoding a colored product, or an enzyme capable of generating products detectable in visible light and therefore measurable. For example, one of the first systems used for reporter gene expression was based on colorimetric assays involving the enzyme β-galactosidase (β-gal) encoded by the gene *lac*Z in the lac operon in *Escherichia coli*. Cleavage of galactosidic substrates by this enzyme generates colored products whose concentration can be measured in a spectrophotometer and is proportional to the amount of enzyme. Thus, the expression of the genes of interest under the promoter of the lac operon can be tracked by monitoring the expression of the β-gal gene. Advantages of this method include its simplicity and low cost, in addition to the vast knowledge about the parameters that could influence the results. Among the main disadvantages are sharing of a promoter, low precision and low reproducibility.

Green Fluorescent Protein (GFP)

The production and detection of fluorescent proteins is already a practical and well characterized technology for sensing and monitoring protein expression of genetic circuits. Furthermore, this technology is not specific and can be applied to a variety of circuits. The cloning of the GFP gene from jellyfish (*Aequorea victoria*) in heterologous systems allows monitoring of its expression via the emission of green fluorescence by the protein after excitation by blue or UV light. For this, scientists couple the gene coding for this protein to the gene of interest using an expression systems and, thus, are able to quantify it and even to locate it in various systems. However, because it is a protein, intrinsic and extrinsic factors involved in its stability, conformation and interactions with various components of the cell may interfere with the accuracy of the methodology. The main advantage of using this protein is the possibility of in vivo and in real time monitoring, without the need for extraction steps or destruction of the particle.

Luminescence

The systems for monitoring protein expression via luminescence emission are based on exothermic reactions that emit luminescence. The main example is undoubtedly those that uses the enzyme luciferase expressed by the *luc* gene, cloned from the genome of firefly (*Photinus pyralis*). This enzyme catalyzes the oxidation of luciferin in the presence of ATP, Mg^{2+} and O_2, generating a light signal that is proportional to the amount of the enzyme. Systems based on luminescence present high sensitivity, speed and are relatively inexpensive. However, in the same way using GFPs, the results depend on the efficiency of the maintenance of the enzymatic activity by stabilizing the system conditions. Specifically, these system also present a great sensitivity to variations in ATP concentrations, which can make their use unfeasible in some cases.

Aptamers

Aptamers are fragments of macromolecules, specially nucleic acids, that are capable of specifically recognizing and detecting proteins. However, their in vivo use has not been widely applied to the monitoring of genetic circuits because the algorithms that could safely predict the binding affinity between aptamers and the proteins of interest would be extremely specific and therefore not viable, since they are not applied to different cases. However, alternatives that use these molecules also can be developed for detection of proteins in several different systems.

Detection of Cellular mRNAs

The amplification of the mRNA for quantitation of the expression of certain genes constitutes a simple and well-characterized technique for monitoring the expression of genes contained in genetic circuits and governed by different modulators. However, besides not being an accurate picture of the protein content of the host, thus excluding the translational stages, techniques for detecting nucleic acids require the extraction of this material and consequently the destruction of the cells analyzed. This is therefore an invasive technique and impractical for real time analyses. The detection of RNAs by fluorescent proteic probes can be an alternative to these issues, but requires that the RNA of interest are modified so that it can specifically bind to the probe.

Debugging

The number of genetic circuits that have presented unexpected results is still very high. This is due to our still limited knowledge about the different parameters involving biological circuits for the synthesis of viral genomes, and the interactions between the components of these circuits and the machinery of the host. In addition, living organisms do not respond linearly according to its components due to the co-operability between molecules, which makes even more challenging to develop predictive mathematical models for these systems. The debugging of genetic circuits involves the detailed characterization of these systems, all available nodes, modifications by trial and error to characterize the interactions between the molecules involved, and specific modeling.

Examples of Projects on Synthetic Genomes

1. *Bacterial gene circuits:* In 2004, the group of researchers leaded by Hideki Kobayashi and colleagues engineered genetic circuits coupled to the cellular machinery of an *E. coli* strain. For this, they inserted plasmids containing a gene essential for biofilm formation under the control of a repressor regulated by RecA protein, which is responsible for mediating the SOS response. The circuit designed was capable of responding to biological signals (damage to genomic DNA) in a predictable and programmable way, so that cells formed biofilm when stimulated. With this, the researchers showed that the genetic circuits can be used to program bacterial cells, through coupling these circuits to the interactions networks of the cell itself (Kobayashi et al. 2004).
2. *Synthetic viral genomes:* Aiming the degradation of bacterial biofilms, American researchers engineered the genome of bacteriophage T7, an *E. coli*-specific phage, in which they inserted the gene for the enzyme dispersin B, which is able to reduce biofilms of different species of bacteria. The gene was inserted under the control of the T7phi10 promoter, so that the enzyme was expressed during infection of the phage and thus released after bacterial lysis. The engineered phage was about two orders of magnitude more efficient in degrading the biofilm matrix as compared to the wild phage (T K Lu & J J Collins 2007).
3. *Synthetic viral genomes:* using tools of synthetic biology, Paul R. Jaschke and colleagues recently published a study aiming establish methods enabling the synthesis, assembly, and recovery of a

bacteriophage genome via yeast. They used the bacteriophage øX174 given its small circular genome with 5386 nucleotides that codes 11 gene products with overlapping open reading frames. The researchers questioned whether this overlap was essential or might be eliminated. For this, they synthesized a 6302 nucleotide synthetic genome separating all ORFs and their regions of translational control. The synthesis of the synthetic genome was performed by a combination of two different strategies: chemical synthesis and PCR amplification. For the synthetic genome was reduced to the exact size of the wild genome, the F gene, which encodes the coat protein, was truncated. The researchers were able to recover infective virions containing the truncated and decompressed genome from replication cycles in *E. coli* expressing the F gene under an inducible promoter. These results indicate that the overlap of genes in the genome of this phage is probably just a tool for genome compaction to facilitate its packaging into the viral particles, but is not necessary for the virions viability (Jaschke et al. 2012).

Prospects

Although gene manipulation and processing of viral genomes are already routinely used in laboratories for the modification of viral characteristics, the development of entirely new synthetic genomes without the use of templates and considering the genetic circuits as an unit for assembly of genomes, comprises a new theme and is still relatively obscure. The hurdles to predict and forecast the synergistic behavior of these circuits lead to security issues that must be answered before this method becomes popular. This only will be possible from the knowledge on the interactions between the different components of this circuit, and the monitoring and debugging of the already created circuits in order to increase the existing mathematical models and algorithms to optimize the predictability of these systems. Even so, the creation of entirely new viral genomes is considered one of the most promising technologies for the development of new methods of genetic transfer and recombination, for the generation of new more specific and efficient antimicrobials, for the planning of new vaccines and specific antiviral drugs with lower side effects, and also for the detection of microorganisms in various industrial and hospital environments, where strict control of the microbial population is essential.

Conflict of Interest Statement

None of the authors of this paper has a financial or personal relationship with other people or organizations that could inappropriately influence or bias the content of this chapter.

Acknowledgment

This work was supported by Brazilian funding agencies CNPq and FAPEMIG.

References

Abrescia, N. G. et al. (2012). Structure unifies the viral universe. *Annual review of biochemistry*, *81*, 795–822. Available at: http://www.ncbi.nlm.nih.gov/pubmed/22482909.

Botstein, D. (1980). A theory of modular evolution for bacteriophages. *Annals of the New York Academy of Sciences*, *354*, 484–490. Available at: http://www.ncbi.nlm.nih.gov/pubmed/6452848.

Chen, Y. S. et al. (2013). DNA sequencing using electrical conductance measurements of a DNA polymerase. *Nature nanotechnology*. Available at: http://www.ncbi.nlm.nih.gov/pubmed/23644569.

Cheung, A. K. (2012). Porcine circovirus: transcription and DNA replication. *Virus research*, *164*(1-2), 46–53. Available at: http://www.ncbi.nlm.nih.gov/pubmed/22036834.

Claverie, J. M. & Abergel, C. (2010). Mimivirus: the emerging paradox of quasi-autonomous viruses. *Trends in genetics : TIG*, *26*(10), 431–437. Available at: http://www.ncbi.nlm.nih.gov/pubmed/20696492.

Dolja, V. V & Koonin, E. V. (2011). Common origins and host-dependent diversity of plant and animal viromes. *Current opinion in virology*, *1*(5), 322–331. Available at: http://www.ncbi.nlm.nih.gov/pubmed/ 22408703.

Eisenstein, M. (2012). Oxford Nanopore announcement sets sequencing sector abuzz. *Nature biotechnology*, *30*(4), 295–296. Available at: http://www.ncbi.nlm.nih.gov/pubmed/22491260.

Ferber, D. (2004). Synthetic biology. Microbes made to order. *Science (New York, N.Y.), 303*(5655), 158–61. Available at: http://www.ncbi.nlm.nih.gov/pubmed/14715986 [Accessed March 11, 2013].

Gibson, D. G. et al. (2010). Creation of a bacterial cell controlled by a chemically synthesized genome. *Science (New York, N.Y.), 329*(5987), 52–6. Available at: http://www.ncbi.nlm.nih.gov/pubmed/20488990 [Accessed February 27, 2013].

Hunkapiller, T. et al. (1991). Large-scale and automated DNA sequence determination. *Science, 254*(5028), 59–67. Available at: http://www.ncbi.nlm.nih.gov/pubmed/1925562.

Jaschke, P. R. et al. (2012). A fully decompressed synthetic bacteriophage øX174 genome assembled and archived in yeast. *Virology, 434*(2), 278–84. Available at: http://www.ncbi.nlm.nih.gov/pubmed/23079106 [Accessed May 15, 2013].

Kobayashi, H. et al. (2004). Programmable cells: interfacing natural and engineered gene networks. *Proceedings of the National Academy of Sciences of the United States of America, 101*(22), 8414–9. Available at: http://www.pubmedcentral.nih.gov/articlerender.fcgi?artid=420408&tool=pmcentrez&rendertype=abstract [Accessed March 18, 2013].

Kristensen, D. M. et al. (2010). New dimensions of the virus world discovered through metagenomics. *Trends in microbiology, 18*(1), 11–19. Available at: http://www.ncbi.nlm.nih.gov/pubmed/19942437.

Lawrence, J. G., Hatfull, G. F. & Hendrix, R. W. (2002). Imbroglios of viral taxonomy: genetic exchange and failings of phenetic approaches. *Journal of bacteriology, 184*(17), 4891–4905. Available at: http://www.ncbi.nlm.nih.gov/pubmed/12169615.

Lima-Mendez, G., Toussaint, A. & Leplae, R. (2011). A modular view of the bacteriophage genomic space: identification of host and lifestyle marker modules. *Research in microbiology, 162*(8), 737–746. Available at: http://www.ncbi.nlm.nih.gov/pubmed/21767638.

Little, J. W. (2010). Evolution of complex gene regulatory circuits by addition of refinements. *Current biology : CB, 20*(17), R724–34. Available at: http://www.ncbi.nlm.nih.gov/pubmed/20833317.

Liu, J., Chen, I. & Kwang, J. (2005). Characterization of a previously unidentified viral protein in porcine circovirus type 2-infected cells and its role in virus-induced apoptosis. *Journal of virology, 79*(13), 8262–74. Available at: http://www.pubmedcentral.nih.gov/articlerender.fcgi?artid=1143760&tool=pmcentrez&rendertype=abstract [Accessed May 17, 2013].

Lu, T. K. & Collins, J. J. (2007). Dispersing biofilms with engineered enzymatic bacteriophage. *Proceedings of the National Academy of Sciences of the United States of America*, *104*(27), 11197–11202. Available at: http://www.ncbi.nlm.nih.gov/pubmed/17592147.

Lu, T. K., Khalil, A. S. & Collins, J. J. (2009). Next-generation synthetic gene networks. *Nature biotechnology*, *27*(12), 1139–1150. Available at: http://www.ncbi.nlm.nih.gov/pubmed/20010597.

Lu, Timothy K. (2010). Engineering scalable biological systems. *Bioengineered bugs*, *1*(6), 378–84. Available at: http://www.pubmedcentral.nih.gov/articlerender.fcgi?artid=3056087&tool=pmcentrez&rendertype=abstract [Accessed April 3, 2013].

Malet, H. et al. (2008). The flavivirus polymerase as a target for drug discovery. *Antiviral research*, *80*(1), 23–35. Available at: http://www.ncbi.nlm.nih.gov/pubmed/18611413.

Marinelli, L. J., Hatfull, G. F. & Piuri, M. (2012). Recombineering: A powerful tool for modification of bacteriophage genomes. *Bacteriophage*, *2*(1), 5–14. Available at: http://www.ncbi.nlm.nih.gov/pubmed/22666652.

McNair, K., Bailey, B. A. & Edwards, R. A. (2012). PHACTS, a computational approach to classifying the lifestyle of phages. *Bioinformatics*, *28*(5), 614–618. Available at: http://www.ncbi.nlm.nih.gov/pubmed/22238260.

Pellegrini, M. et al. (1999). Assigning protein functions by comparative genome analysis: protein phylogenetic profiles. *Proceedings of the National Academy of Sciences of the United States of America*, *96*(8), 4285–4288. Available at: http://www.ncbi.nlm.nih.gov/pubmed/10200254.

Raoult, D. et al. (2004). The 1.2-Mb Genome Sequence of Mimivirus. *Science*, *306*, 1344-1350. Available at: http://www.ncbi.nlm.nih.gov/pubmed/15486256 [Accessed May 24, 2013].

Radford, A. D. et al. (2012). Application of next-generation sequencing technologies in virology. *The Journal of general virology*, *93*(Pt 9), 1853–1868. Available at: http://www.ncbi.nlm.nih.gov/pubmed/22647373.

Riccione, K. A. et al. (2012). A synthetic biology approach to understanding cellular information processing. *ACS synthetic biology*, *1*(9), 389–402. Available at: http://www.ncbi.nlm.nih.gov/pubmed/23411668 [Accessed March 7, 2013].

Rowen, L., Mahairas, G. & Hood, L. (1997). Sequencing the human genome. *Science*, *278*(5338), 605–607. Available at: http://www.ncbi.nlm.nih.gov/pubmed/9381170.

Stéphane Leduc (1910). *Théorie Physico-Chimique de la Vie et Générations Spontanées*, 1st ed. A. Poinat Editer, ed., Paris.

Toussaint, A., Lima-Mendez, G. & Leplae, R. (2007). PhiGO, a phage ontology associated with the ACLAME database. *Research in microbiology*, *158*(7), 567–571. Available at: http://www.ncbi.nlm.nih.gov/pubmed/17614261.

Uetz, P. et al. (2000). A comprehensive analysis of protein-protein interactions in Saccharomyces cerevisiae. *Nature*, *403*(6770), 623–627. Available at: http://www.ncbi.nlm.nih.gov/pubmed/10688190.

Veesler, D. & Cambillau, C. (2011). A common evolutionary origin for tailed-bacteriophage functional modules and bacterial machineries. *Microbiology and molecular biology reviews : MMBR*, *75*(3), 423–33, first page of table of contents. Available at: http://www.ncbi.nlm.nih.gov/pubmed/21885679.

Westerhoff, H. V. & Palsson, B. O. (2004). The evolution of molecular biology into systems biology. *Nature biotechnology*, *22*(10), 1249–1252. Available at: http://www.ncbi.nlm.nih.gov/pubmed/15470464.

Yu, D. et al. (2000). An efficient recombination system for chromosome engineering in Escherichia coli. *Proceedings of the National Academy of Sciences of the United States of America*, *97*(11), 5978–5983. Available at: http://www.ncbi.nlm.nih.gov/pubmed/10811905.

Zhang, J. et al. (2011). The impact of next-generation sequencing on genomics. *Journal of genetics and genomics = Yi chuan xue bao*, *38*(3), 95–109. Available at: http://www.pubmedcentral.nih.gov/articlerender.fcgi?artid=3076108&tool=pmcentrez&rendertype=abstract [Accessed February 28, 2013].

In: Advances in Viral Genomes Research
Editors: J. Borrelli and Y. Giannini
ISBN: 978-1-62808-723-9

Chapter 4

Novel Bioinformatics Method to Analyze More than 10,000 Influenza Virus Strains Easily at Once: Batch-Learning Self Organizing Map (BLSOM)

Yuki Iwasaki[1,2], Toshimichi Ikemura[1], Kennosuke Wada[1], Yoshiko Wada[1,3] and Takashi Abe[1,4,*]

[1]Nagahama Institute of Bio-Science and Technology,
[2]Research Fellow of the Japan Society for the Promotion of Science,
[3]Shiga University of Medical Science,
[4]Niigata University, Japan

Abstract

With increase of microbial and viral genome sequence data obtained from high-throughput DNA sequencers, novel tools are needed for comprehensive analyses of the big sequences data. An unsupervised neural network algorithm, Self-Organizing Map (SOM), is an effective tool for clustering and visualizing high-dimensional complex data on a

[*] Corresponding author: Takashi Abe. Email: takaabe@ie.niigata-u.ac.jp.

single map. We previously modified the conventional SOM for genome informatics on the basis of batch-learning SOM (BLSOM), by making the learning process and resulting map independent of the order of data input. Influenza virus is one of zoonotic viruses and shows clear host tropism. Important issues for bioinformatics studies of influenza viruses are prediction of genomic sequence changes in the near future and surveillance of potentially hazardous strains.

To characterize sequence changes of influenza virus genomes after invasion into humans from other animal hosts and to study molecular evolutionary processes of their host adaptation, we have constructed BLSOMs for oligonucleotide, codon, amino-acid, and peptide compositions in all genome sequences of influenza A and B viruses and found clear host-dependent clustering (self-organization) of the sequences.

Viruses isolated from humans and birds differ in mononucleotide composition from each other. In addition, host-dependent oligonucleotide and peptide compositions that cannot be explained with the host-dependent mononucleotide composition are revealed by these BLSOMs. Retrospective time-dependent directional changes of oligonucleotide compositions, which are visualized for human strains on BLSOMs, can provide predictive information about sequence changes of the newly invaded viruses from other animal sources.

Basing on this host-dependent oligonucleotide composition, we have proposed a strategy for prediction of directional changes of virus sequences and for surveillance of potentially hazardous strains when introduced into human populations from nonhuman sources. Millions of genomic sequences from infectious microbes and viruses will become available in the near future because of their medical importance, and BLSOM can characterize such big data easily and support efficient knowledge discovery.

Introduction

Due to the revolutionized advancement of decoding the genome sequences, their data have been growing into supermassive "big data" in the International DNA Data Banks, and therefore, large-scale data mining have become vital. The more the available information becomes larger, the more the rightfulness of the proposed model will become verifiable as long as the model is appropriate. At the same time, because of massive amounts of data available, analyzing only a limited part of the data should lead to an image of "the authors may have made convenient stories by extracting useful data for them", especially when a novel discovery is publicized. It should be important to have a stance "all data belonging to a certain category have to be analyzed, even if the amount of the data is very large". Therefore, in the post-genome era, the research that gets the picture of the whole will become increasingly important and the present chapter introduces a bioinformatics method suitable for this type of a large-scale analysis.

In more detail, we introduce the studies to unveil the species-specific characteristics in the genome sequences by using "BLSOM" developed by our group [1, 2]. This BLSOM for oligonucleotide composition has been developed for genome informatics, on the basis of SOM (Self-Organization Map) originally established by Kohonen and his colleagues [3, 4]. Because BLSOM can analyze millions of sequences simultaneously, almost all known genome sequences have been clustered and analyzable on a single map [5].

GC% has been used for a long period as a fundamental parameter for phylogenetic classification of microbial genomes, including viral genomes, but the GC% is apparently too simple a parameter to differentiate a wide variety of microbial genomes. Oligonucleotide composition, on the other hand, can be used to distinguish the species even with the same GC%, because the oligonucleotide composition varies significantly among the genomes and is called 'genome signature' [6]. Previously, we have constructed BLSOMs for di-, tri-, or tetranucleotide composition in all genomic fragments (e.g. 10, 50 and 100 kb) derived from a wide range of prokaryotic and eukaryotic species and found that, without giving any information of species, most of the fragment sequence have been separated (self-organized) according to species [1, 2, 5].

Epidemics of influenza viruses have been frequently repeated among various animal species including birds and swine besides human. Given that the strains derived from nonhuman sources are also capable of infecting humans, it should be impossible to eradicate influenza viruses from the planet. Influenza virus has developed into a pandemic among humans at least four times (in 1918, 1957, 1968 and 2009), causing considerable damages. Virus requires many host factors (e.g. nucleotide, amino acid and tRNA pools and host proteins) when it grows; therefore, it inescapably depends on new host factors when causing a new infection to humans from other animal species. In such cases, the genome sequence of the virus will change to fit the new cellular environments including various host factors. If this view is correct, influenza viruses are likely to modify some part of the characteristics of their genome sequences depending on host, during epidemics in a new host. In other words, it enables us to predict, at least in part, the direction of sequence changes in the newly invaded virus, if we can properly find out the host-dependent characteristics of viral genome sequences. In this chapter we introduce examples of analysis on the host-dependent characteristics of influenza virus genome sequences, which is a key to conduct forecasting their sequence changes after changing hosts. In the previous study, we have attempted BLSOM analyses on viral genome sequences all available in that time [7]. Here, we introduce the BLSOM analysis, which includes the newly accumulated sequences and examine whether the directional changes proposed in the previous paper is actually observed in the sequences obtained after our previous paper. This type of confirmatory studies should be important to verify a novel bioinformatics method.

Materials and Methods

Viral Genome Sequences

A total of 856,730 virus sequences analyzed in Figure 1 were obtained from the NCBI and a total of 12,370 influenza A and B virus strains analyzed in Figure 2 were obtained from the NCBI Influenza Virus Resource (http://www. ncbi.nlm.nih.gov/genomes/FLU/) [8].

Batch-Learning Self-Organizing Map (BLSOM)

SOM is an unsupervised neural network algorithm that implements a characteristic non-linear projection from the high-dimensional space of input data onto a two-dimensional plane [3, 4]. We modified the conventional SOM for genome informatics to make the learning process and resulting map independent of the order of data input on a basis of Batch-Learning SOM: BLSOM [1, 2, 9]. The initial weight vectors were defined by principal component analysis instead of random values. BLSOM learning for oligonucleotide composition was conducted as described previously [1,2]. BLSOM learning for synonymous codon usage and visualization of diagnostic codons for the category separation were conducted as described by Kanaya et al. [9]. BLSOM program was obtained from UNTROD, Inc. (y_wada@ nagahama-i-bio.ac.jp).

Oligonucleotide composition is normalized for the sequence length and the normalized composition is used for BLSOM calculation. Mono-, di- or tri-amino acid composition in eight genes of influenza virus strains is calculated, and those of eight genes are summed up for each strain. The compostion normalized with a total length is used for BLSOM calculation.

Result

Oligonucleotide BLSOM for All Virus Genome Sequences

To explain the clustering powers of BLSOM for a large number of genome sequences from a wide variety of viruses, we have initially constructed BLSOM for di- and trinucleotide composition in all viral genome sequences currently available (ca. 860,000 sequences), which have been classified into 67 phylogenic families (Figure 1). Lattice points that contain sequences from one phylogenetic family are indicated in color, and those that include sequences from more than one family are indicated in black. A major portion is colored, which shows that a major portion of sequences is classified (self-organized) according to phylogenic family; 95 and 98% of viral sequences are located in colored lattice points on di- and tri-BLSOMs, respectively. Notably, no information in regard to phylotype has been given during the BLSOM calculation. It should also be mentioned that approximately one million sequences are analyzed simultaneously on one plane.

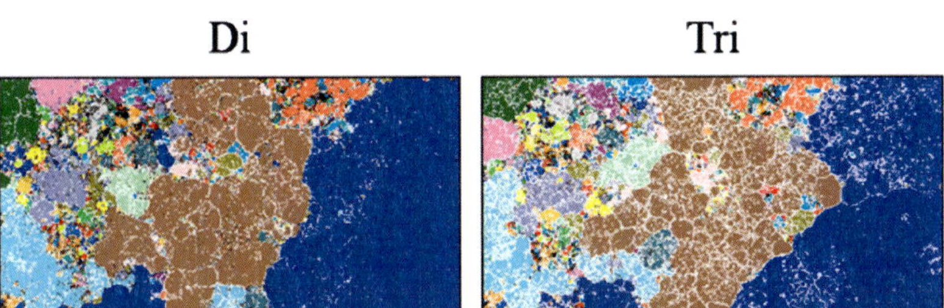

Figure 1. Oligonucleotide-BLSOMs for virus genome sequences. BLSOM has been constructed for di- or trinucleotide composition (Di or Tri) in 856,730 viral genome sequences. Lattice points that include sequences from more than one phylogenetic family are indicated in black, and those containing sequences from a single phylotype are indicated in a color representing the phylotype: Arteriviridae (■), Bunyaviridae (■), Coronaviridae (■), Flaviviridae (■), Geminiviridae (■), Hepadnaviridae (■), Herpesviridae (■), Orthomyxoviridae (■), Paramyxoviridae (■), Picornaviridae (■), Potyviridae (■), Reoviridae (■), Retroviridae (■), Rhabdoviridae (■) and others (various colors not specified here).

Oligonucleotide BLSOM for Influenza Virus Genomes

Over ten thousand genome sequences of influenza A and B virus strains are registered in the International Nucleotide Sequence Databases, even when we concern strains for which all eight segments have been sequenced. We have next constructed a tetranucleotide BLSOM for all influenza A or B strains registered in Influenza Virus Resource [8] in NCBI (Figure 2); for di- and trinucleotide BLSOMs, refer to our previous paper [7]. Because the direct target of natural selection is a virion containing a full set of eight genome segments, tetranucleotide frequencies in the eight segments are summed up for each strain and BLSOM is constructed with the summed frequency after normalization of a total nucleotide length of individual strains; this is because the length of genome sequences compiled by the Influenza Virus Resource differs slightly between strains (Figure 2). The present genome-level analysis should provide valuable information for characterizing individual strains, which may not be obtained by a gene-level analysis, primarily conducted with sequence homology searches used in phylogenetic tree analyses.

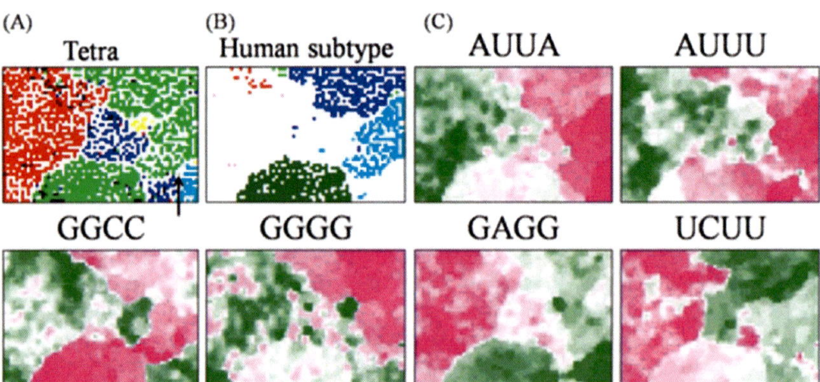

Figure 2. Tetranucleotide-BLSOM for influenza A and B virus genome sequences. (A) Tetranucleotide BLSOM (Tetra) has been constructed for 12,370 strains of influenza A and B virus strains. Lattice points containing sequences from strains isolated from more than one host are indicated in black, and those containing sequences from one host are indicated in a color: avian, red; human, green; swine, blue; equine, yellow; bat, grey; and influenza B strain, light blue. (B) Human subtype. On the tetra-BLSOM presented in (A), each human virus subtype is specified in a color: H1N1, light blue; H1N1/09, dark green; H3N2, blue; H5N1, red; H7 and H9, pink; and other human subtype, green. (C) Occurrence levels of tetranucleotides, which are diagnostic for host-dependent clustering, are indicated with different levels of two colors: pink (high), green (low), and achromatic (intermediate) as described by Abe et al. [1].

Lattice points containing virus strains isolated from one host species are indicated in a color representing the host and those containing strains isolated from more than one host are in black. Though only oligonucleotide frequencies have been given during the BLSOM calculation, viral sequences are clustered (self-organized) according to host. Viruses are inevitably dependent on many host factors for their growth (e.g. pools of nucleotides, amino acids and tRNAs) and at the same time have to escape from antiviral host mechanisms such as antibodies, cytotoxic T cells, interferons, and RNA interferences [10-13]. Host-dependent clustering of viral sequences visualized on BLSOM should reflect this host dependency in their growth.

In Figure 2B (Human Subtype), lattice points that contain human influenza A viruses of one subtype on the tetra-BLSOM (Figure 2A) are specified with one color representing the subtype. Among the 12,000 viral strains analyzed, approximately 3,000 strains correspond to the new pandemic H1N1 strain (H1N1/09) (dark green in Figure 2B), which has started its pandemics among humans around since April 2009. Although its origin is derived from avian strains, it has been infected to human via swine after genome segment reassortments. Interestingly, on BLSOM, they are located apart from seasonal human H1N1 or H3N2 strains (light blue or dark blue in Figure 2B) and surrounded by avian and swine territories (red and blue in Figure 2A and blank in Figure 2B). It can be possible that these H1N1/09 strains have not yet been best suited to growth under human cellular environments.

We next investigate human virus subtypes other than the H1N1/09 (Figure 2B). Human seasonal H1N1 and H3N2 are clearly separated from each other; light and dark blue in Figure 2B. In addition, in contrast to H1N1/09 (dark green), human H5N1 strains (red in Figure 2B and mainly black in Figure 2A) are rather scattered within the avian territory (red in Figure 2A and blank in Figure 2B). This should show that the human H5N1 strains have jumped to humans but are not able to spread from human to human [12], and therefore, they have characteristics of avian viruses. These human H5N1 strains are more separated from the swine and human territories than H1N1/09 strains, and this difference may relate to their infection power in the human and swine populations.

In the case of influenza B strains (light blue in Figure 2A and blank in Figure 2B), which have repeatedly caused epidemics only among humans, they form a territory more distant from the avian territory than that of human seasonal strains. This characteristic of influenza B strains will be discussed later.

Diagnostic Tetranucleotides Responsible for Host-Dependent Separation

BLSOM provides a powerful ability to visualize diagnostic oligonucleotides responsible for the category-dependent clustering (the host-dependent clustering in this case). In order to study the oligonucleotides that may relate to host adaptation, we next identify diagnostic oligonucleotides for host-dependent separation on the tetranucleotide BLSOM. Six examples of diagnostic tetranucleotides for the host- dependent separation are presented in Figure 2C. A clear tendency is apparent; A- and U-rich oligonucleotides are more favored in humans than in avians (e.g. AUUA and AUUU); G- and C-rich oligonucleotides were more favored in avians than in humans (e.g. GAGG). This GC% effect was previously reported by Rabadan et al. [14]. However, GGCC and GGGG, which are composed only of G and/or C, are more favored in humans than in avians, and UCUU, a tetranucleotide rich in U, is preferred mainly in the avian territory. When we concern about interaction of viral RNAs (vRNAs and mRNAs) with various host factors (e.g. host proteins), oligonucleotide compositions, rather than the mononucleotide composition, will become important, because their interactions with host proteins primarily depend on oligonucleotide sequences in their sequences. This should also be true in considering escape processes from host antiviral mechanisms. To wrap up, oligonucleotide-BLSOM analyses of influenza viruses should provide valuable information of their host adaptation mechanisms (e.g. interactions of viral RNAs with host factors), which will become important for medical and pharmaceutical studies.

Chronological Changes Visualized for Human Viruses

In Figure 3, we next explain chronological changes of human seasonal strains using the same tetranucleotide BLSOM; only the lattice points containing human strains are shown as done in Figure 2B. In addition, the territory of human seasonal H1N1 (except for H1N1/09) and H3N2 strains in the following period is separately displayed in brown and blue; H1N1 and/or H3N2 strains isolated in 1930-1957, 1968-1974, 1975-1989, 1990-1999, 2000-2005, and 2006-2012 are displayed in brown and blue, respectively, in the respective panel. The seasonal strains isolated in 1930-1957 and 1968-1974 (i.e. the beginning of the epidemic of H1N1 and H3N2, respectively) are seen near avian and/or swine strains (red and blue in Figure 2A and blank in Figure

3), which have been estimated as their origin. Human seasonal strains are seen moving away from the avian and/or swine territory since 1930 (for H1N1) and 1968 (for H3N2), indicating the directional change of their oligonucleotide composition after invasion into the human population. Therefore, at least from a specific point of view, genome sequence changes in the invader virus appear to be predictable, especially in an early stage of epidemics cycles. This has been found for H1N1/09 strains in our previous paper [7] even during the first one year, because of the high mutation rate and short generation time of influenza A viruses [12, 15].

To wrap up, BLSOM can visualize any categories of strains, in which experimental and medical groups will be interested, and therefore, has a power to visualize evolutionary histories of influenza viruses after invasion into new hosts.

Influenza B has caused repetitive epidemics only among humans and can be considered as it has already been well adapted to human cellular environments. In order to examine numerically whether the tetranucleotide composition of human seasonal A strains are approaching that of Influenza B with time, we have calculated Euclidean distance between the average composition in influenza B strains and that in H1N1 or H3N2 strains isolated in each year (Figure 4A). Euclidean distance shows the distance in the multidimensional scale (256 dimension for tetranucleotide composition); the distance between the strains that own an identical oligonucleotide composition is 0 and, as the difference in the oligonucleotide composition grows, the distance becomes larger.

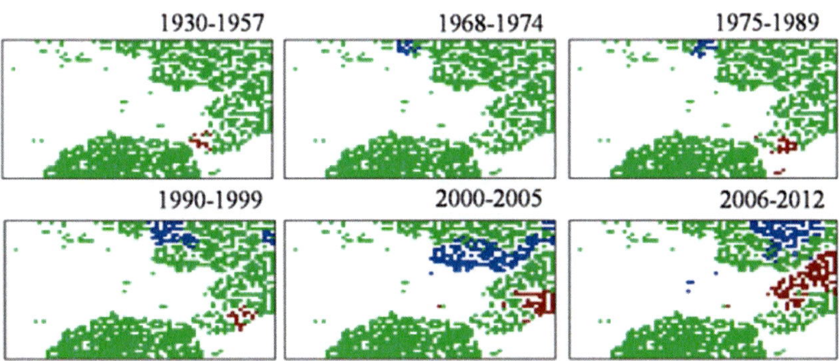

Figure 3. Chronological changes for seasonal human subtype strains. Seasonal human H1N1 and H3N2 strains that are isolated in the different period are separately marked in brown and blue, respectively.

The distance between the influenza B strain and either H1N1 (red + marks) or H3N2 (blue x marks) A strain has become smaller over time (Figure 4A), showing that the tetranucleotide composition in A strains has become analogous to that in B strain during the repeating pandemics among humans and directional changes have been accumulating in genome sequences in the seasonal A strains.

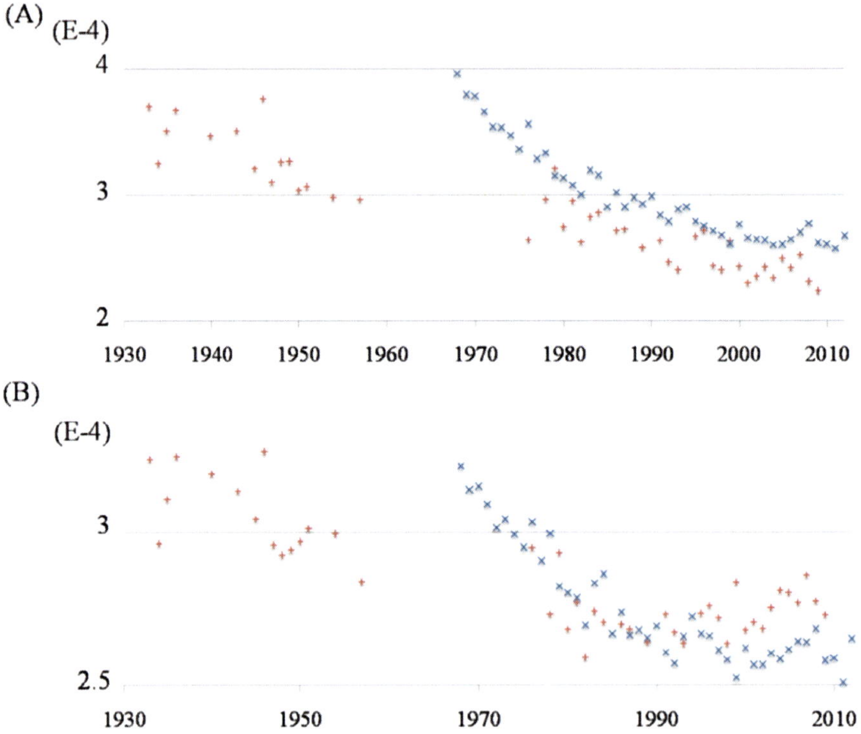

Figure 4. Euclidean distance between seasonal human A and B strains. (A) Euclidean distance of tetranucleotide composition between influenza B strain and human seasonal strain (influenza A strain). Euclidean distance between the average composition in influenza B strains and that in H1N1 (+) or H3N2 (×) strains isolated in each year is plotted. (B) Euclidean distance of tetranucleotide composition between the bat and human seasonal strains. Euclidean distance between the average composition in bat strains and that in H1N1 (+) or H3N2 (×) strains isolated in each year is plotted.

H17N10 Strains Isolated from Bat

One of the most feared influenza viruses is the strain possessing antigen types, to which most humans do not possess antibodies, and therefore, the strains having caused epidemics in nonhuman sources (e.g. birds and pigs) attract attention.

Influenza A viruses are categorized by the combinations of antigen types: 16 and 9 types of HA and NA, respectively. As a natural host, birds are capable of transmitting all of the 16×9=144 types of influenza A strains. During recent years, new H17N10 strains, which greatly vary from known influenza A strain, have been isolated from the bats in Republic of Guatemala and estimated as being separated from birds far into the past [16]. As previously mentioned, BLSOM have a power to focus on and visualize a specific category of strains with no difficulty, even massive amounts of strains are analyzed. When paying attention to the three bat strains on BLSOM (arrowed in Figure 2A), these are located near influenza B strains and thus far away from the territory of birds which should be their original sources. This indicates that these bat H17N10 strains have been well adapted to mammalian cellular environments during repetitive epidemics in the mammalian host. So far judged from locations on BLSOM, cellular environments of human and bat appear to resemble each other, because human influenza B and bat A strains are located in a close vicinity.

Next, this prediction has been numerically checked, as done for the human influenza B strain in Figure 4A. In Figure 4B, we have plotted the Euclidean distance between the average composition in three bat H17N10 strains and that in H1N1 or H3N2 strains isolated in each year. As observed in Figure 4A, both H1N1 and H3N2 strains have shortened their Euclidean distances from the bat strains over time, indicating that cellular environments of human and bat appreciably resemble each other for the influenza virus growth. To sum up the findings in Figure 4A and B, the tetranucleotide composition most suitable to the growth in mammalian cells differs significantly from that for the growth in avian cells, and the avian strains invaded into a new host have to change their sequences for obtaining the composition suitable to the new cellular environments. This directional change can offer information on predictions of viral sequence changes that will occur in the near future.

BLSOM for Codon Usage

Synonymous codon choice sensitively reflects constraints imposed on genome sequences and thus provides a sensitive probe for searching for molecular mechanisms responsible for the constraints, e.g. genome GC% and tRNA composition in the cases of micro-organisms [17-20]. We have previously found that BLSOM efficiently detects species-specific codon-choice patterns of micro-organisms, resulting in self-organization of genes according to microbial species [9]. Furthermore, in the case of genes horizontally transferred relatively recently, synonymous codon choice reflects primarily that of the donor, but not the recipient, genome. In the case of influenza viruses, we have previously found host-dependent clustering on codon-BLSOM [7], and in this chapter we introduce BLSOM for synonymous codon usage in influenza A and B virus genes (Figure 5A), by analyzing sequences including those published after our previous paper. In order to know codon biases for each strain, codon usages in eight genes are summed up for each strain.

Human and avian territories are again clearly separated from each other, and human H1N1/09 strains (dark green in Figure 5B) are again separated from the major human territory and surrounded by avian, equine and swine territories.

Synonymous codon-choice patterns of newly invading viruses, such as H1N1/09, should be close to those of the original host viruses, at least for a period immediately after the invasion. Codon choice will most likely shift towards the pattern of seasonal human viruses during many infection cycles among humans, because viruses depend on many cellular factors for their growth. Actually, the directional sequence changes in H1N1/09 towards the codon usage pattern of the seasonal human strains have been observed in our previous paper [7].

When we focus on diagnostic codons (Figure 5C), one simple tendency is observed. Codons ending with G or C are more favored for the avian strains than the human strains (AAG for Lys in Figure 5C), and vice versa. While the GC% effect is most apparent in two-codon boxes, this is also observed for many codons in four- or six-codon boxes (four examples in Figure 5C).

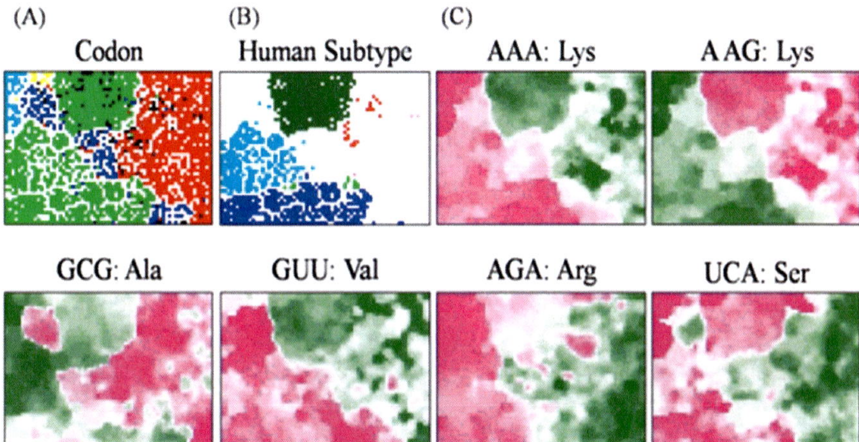

Figure 5. BLSOMs for codon usage. (A) Codon. BLSOM has been constructed for synonymous codon usage in 12,288 genes from influenza A and B strains, and lattice points are indicated in a color representing the host as described in Figure 2A. (B) Human subtype. On the Codon-BLSOM presented in (A), lattice points are indicated in a color representing the host as described in Figure 2B. (C) Occurrence levels of six codons, which are diagnostic for host-dependent separation, are indicated with different levels of two colors, as described in Figure 2C.

BLSOM for Amino-Acid and Peptide Composition

When we consider prediction of changes in virus sequences from a view of medical and pharmaceutical application, changes in amino acid sequences appear to be more informative than in nucleotide sequences, because amino acid sequences are more directly related to viral functions than genomic sequences; e.g. prediction of amino-acid sequence changes is very useful for designing vaccine and antiviral drug.

Therefore, we next examine whether directional change of amino-acid and peptide composition can be detected with BLSOMs. We have constructed BLSOMs for mono amino-acid, dipeptide and tripeptide composition for a sum of major eight proteins (PB2, PB1, PA, HA, NP, NA, M1, and NS1) (Figures 6A and 7A, C).

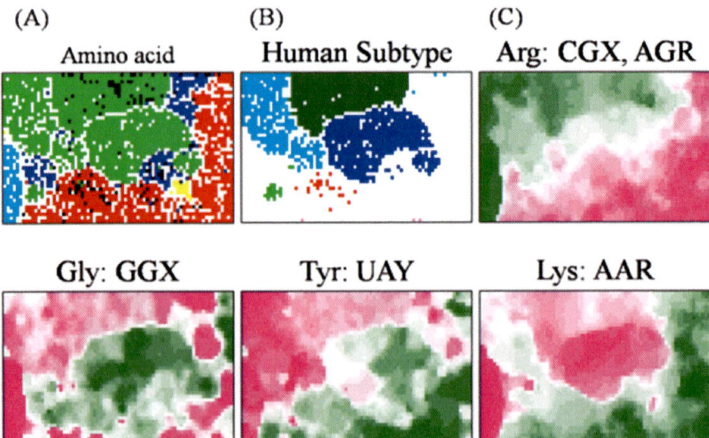

Figure 6. BLSOMs for amino-acid composition. (A) Amino acid. BLSOM has been constructed for amino-acid composition in 12,288 genes from influenza A and B strains, and lattice points are indicated in a color representing the host as described in Figure 2A. (B) Human subtype. On the Amino-BLSOM presented in (A), lattice points are indicated in a color representing the host as described in Figure 2B. (C) Occurrence levels of six amino acids, which are diagnostic for host-dependent separation, are indicated with different levels of two colors, as described in Figure 2C.

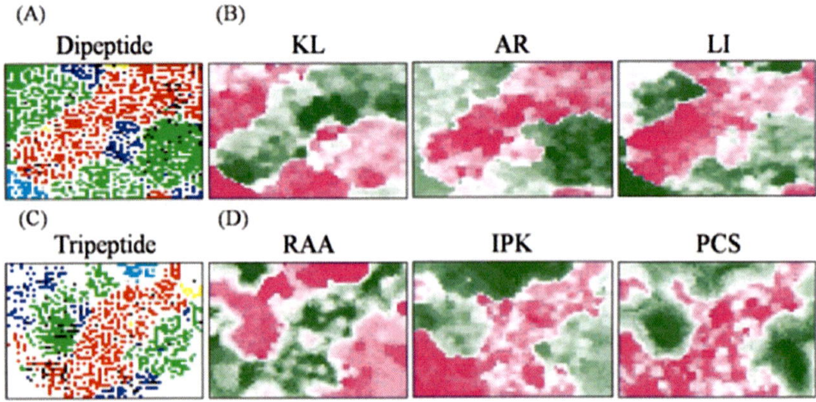

Figure 7. Oligopeptide-BLSOM for all influenza viruses. (A) Dipeptide-BLSOM. The color coding is the same as Figure 2A. (B) Occurrence levels of dipeptide, which are diagnostic for host-dependent separation, are indicated with different levels of two colors, as described in Figure 2C. (C) Tripeptide-BLSOM. The color coding is the same as Figure 2A. (D) Occurrence levels of tripeptide, which are diagnostic for host-dependent separation, are indicated with different levels of two colors, as described Figure 2C.

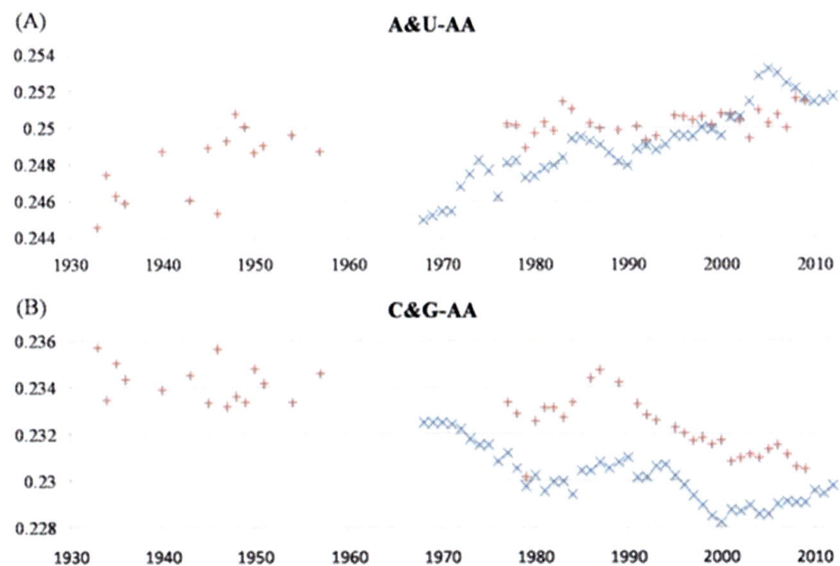

Figure 8. Directional changes of amino-acid frequency in human seasonal strains for A&U-AA and C&G-AA. (A) The A&U-AA frequencies in human seasonal H1N1(+) and H3N2 (×) strains are plotted according to the chronological order. (B) The C&G-AA frequencies in human seasonal H1N1(+) and H3N2 (×) strains are plotted according to the chronological order.

Strains isolated from avian (red) and human (green) are again clustered (self-organized) according to host, forming their own large continuous territories. One simple tendency is immediately apparent. Human strains tend to prefer the amino acids coded by A- and/or U-rich codons (Asn, Ile, Lys, Phe, and Tyr; designated as A&U-AA), while avian strains appear to prefer the amino acids coded by C- and/or G-rich codons (Ala, Gly, Pro, and Arg; designated as C&G-AA) (Figure 6C). This shows that viral protein sequences are strongly affected by the host-dependent base compositions.

In order to examine the directional changes of amino-acid composition during the human virus evolution, we calculate usage frequencies of A&U-AA and C&G-AA separately, and plotted the average of their frequencies in the seasonal human H1N1 or H3N2 strains isolated in each year using the red + or blue x mark (Figure 8). A&U-AAs for both H1N1 and H3N2 strains have been gradually growing since the beginning of their pandemic while C&G-AA appear to be reducing, showing the directional change at the amino-acid level, which is consistent with the direction observed for viral genome GC%.

When we consider functions of individual viral proteins, a combination of amino acids such as di- and tripeptide may become more important than the simple amino-acid composition; i.e. prediction of directional changes of peptide compositions may provide information more directly related to change in functions and/or antigenicities of viral proteins. On both di- and tripeptide BLSOMs, strains isolated from avian (red) and human (green) are again clustered (self-organized) according to host, forming their own large continuous territories (Figure 7A, 7C). Six examples of diagnostic di- or tri-peptide for the host-dependent separation are presented in Figure 7B and 7D; most human strains prefer KL composed only of A&U-AAs and most avian strains prefer AR composed only of C&G-AAs. This is consistent to the prediction from their genome GC%. However, the trend of not relying on viral genome GC% is also seen for various di- and tri-peptides (e.g. RAA in Figure 7B). Host-dependent composition of the latter type of peptides appears to be interesting from a view of their biological significance, and therefore, detailed analyses of these peptides are in progress by our group.

Conclusion

Influenza viruses isolated from humans and birds differed in mono- and oligonucleotide composition and in mono amino-acid and peptide composition. Time-dependent directional changes of these compositions, which are observed for the newly invaded viruses into the human population from other animal hosts, can provide predictive information about sequence changes of the invaded viruses, which should be useful for medical and pharmaceutical purposes. Basing on the findings obtained by BLSOM analyses, we have previously proposed a strategy for efficient surveillance of potentially hazardous nonhuman strains that may cause new pandemics with a high probability [7]. Millions of influenza and other virus sequences will become available in the near future because of their medical and social importance, and BLSOM can characterize such big data without difficulty and support efficient knowledge discovery.

Acknowledgments

This work was supported by Research Fellow of the Japan Society for the Promotion of Science, the Grant-in-Aid for JSPS Fellows (JSPS KAKENHI Number 24•9979), and Grant-in-Aid for Scientific Research (C) and for Young Scientists (B) from the Ministry of Education, Culture, Sports, Science and Technology of Japan. We wish to thank Drs Masae Itoh and Kouki Yonezawa (Nagahama Institute of Bio-Science and Technology) and Dr Kimihito Ito (the Research Center for Zoonosis Control, Hokkaido University) for valuable suggestions and discussions and to Ms. Yukari Yonemori for manuscript preparation.

Reference

[1] Abe, T., Kanaya, S., Kinouchi, M., et al. (2003) Informatics for unveiling hidden genome signatures, *Genome Res.,* 13, 693–702.

[2] Abe, T., Sugawara, H., Kinouchi, M., Kanaya, S. and Ikemura, T. (2005) Novel phylogenetic studies of genomic sequence fragments derived from uncultured microbe mixtures in environmental and clinical samples, *DNA Res.,* 12, 281–90.

[3] Kohonen, T. (1982) Self-organized formation of topologi- cally correct feature maps, *Biol. Cybern.*, 43, 59–69.

[4] Kohonen, T., Oja, E., Simula, O., Visa, A. and Kangas, J. (1996) Engineering applications of the self-organizing map, *Proc. IEEE,* 84, 1358–84.

[5] Abe, T., Sugawara, H., Kanaya, S. and Ikemura, T., Sequences from almost all prokaryotic, eukaryotic, and viral genomes available could be classified according to genomes on a large-scale Self-Organizing Map constructed with the Earth Simulator, *Journal of the Earth Simulator*, 6, 17-23, 2006.

[6] Karlin, S., Campbell, A.M. and Mrazek, J. (1998) Comparative DNA analysis across diverse genomes, *Annu. Rev. Genet.,* 32, 185–225.

[7] Iwasaki, Y., Abe, T., Wada, K., Itoh, M., and Ikemura, T. (2011), Prediction of Directional Changes of Influenza A Virus Genome Sequences with Emphasis on Pandemic H1N1/09 as a Model Case, *DNA res.,* 18, 125-136.

[8] Bao, Y. (2008) The influenza virus resource at the National Center for Biotechnology Information, *J. Virol.,* 82, 596 – 601.

[9] Kanaya, S., Kinouchi, M., Abe, T., et al. (2001) Analysis of codon usage diversity of bacterial genes with a self-organizing map (SOM): characterization of horizontally transferred genes with emphasis on the E. coli O157 genome. *Gene,* 276, 89-99.

[10] García-Sastre, A. (2001) Inhibition of interferon-mediated antiviral responses by influenza A viruses and other negative-strand RNA viruses. *Virology*, 279, 375-384.

[11] Voinnet, O. (2005) Induction and suppression of RNA silencing: insights from viral infections. *Nat. Rev. Genet.*, 6, 206-220.

[12] Nelson, M.I., and Holmes, E.C. (2007) The evolution of epidemic influenza. *Nat. Rev. Genet.*, 8, 196-205.

[13] Alexey, A., and Moelling, K. (2007) Dicer is involved in protection against influenza A virus infection. *J. Gen. Virol.,* 88, 2627-2635.

[14] Rabadan, R., Levine, A.J., and Robins, H. (2006) Comparison of avian and human influenza A viruses reveals a mutational bias on the viral genomes. *J. Virol.,* 80, 11887-11891.

[15] Domingo, E., and Holland, J.J. (1997) RNA virus mutations and fitness for survival. *Annu. Rev. Microbiol.*, 51, 151-178.

[16] Tong, S., Li, Y., Rivailler, P., Conrardy, C., Castillo, DA., Chen, LM., et al. (2012) A distinct lineage of influenza A virus from bats. *PNAS*, 109, 4269-4274.

[17] Ikemura, T. (1981) Correlation between the abundance of Escherichia coli transfer RNAs and the occurrence of the respective codons in its protein genes. *J. Mol. Biol.*, 146, 1-21.

[18] Ikemura, T. (1985) Codon usage and transfer RNA content in unicellular and multicellular organisms. Mol. Biol. Evol., 2, 13-34.

[19] Sharp, P.M., and Matassi, G. 1994, Codon usage and genome evolution. *Curr. Opin. Gen. Dev.,* 4, 851-860.

[20] Sueoka, N. (1995) Intrastrand parity rules of DNA base composition and usage biases of synonymous codons. *J. Mol. Evol.,* 40, 318-325.

Index

A

B

C

D

E

F

G

Q

R

S

T

U

V

W

Y